SpringerBriefs in Applied Statistics and Econometrics

Editors-in-Chief

Michael Massmann, Chair of Econometrics and Statistics, WHU - Otto Beisheim School of Management, Vallendar, Germany

Ostap Okhrin, Chair of Econometrics and Statistics, esp. in Transport Sector, Technische Universität Dresden, Dresden, Germany

Series Editors

Peter Boswijk, Faculty of Economics and Business, University of Amsterdam, Amsterdam, The Netherlands

Jörg Breitung, Intstitute of Econometrics and Statistics, University of Cologne, Köln, Germany

Walter Krämer, Faculty of Statistics, TU Dortmund University, Dortmund, Germany

Karl Mosler, Institute of Econometrics and Statistics, University of Cologne, Köln, Germany

Peter Song, Department of Biostatistics, University of Michigan, Ann Arbor, MI, USA

SpringerBriefs present concise summaries of cutting-edge research and practical applications across a wide range of fields. Featuring compact volumes of 50 to 125 pages, the series covers a range of content, from professional to academic. Briefs are characterized by fast, global electronic dissemination, standard publishing contracts, standardized manuscript preparation and formatting guidelines, and expedited production schedules.

SpringerBriefs in Applied Statistics and Econometrics (SBASE) showcase areas of current relevance in the fields of statistical methodology, theoretical and empirical econometrics, statistics in the natural sciences, and financial econometrics. Of interest are topics arising in the analysis of cross-sectional, panel, time-series and high-frequency data. The primary audience of the SBASE series are researchers with an interest in theoretical, empirical and computer-based statistics and econometrics.

The SBASE series accepts a variety of formats:

- timely reports of state-of-the art techniques,
- bridges between new research results, as published in journal articles, and literature reviews,
- snapshots of a hot or emerging topic,
- lecture or seminar notes making a specialist topic accessible for non-specialist readers.

Manuscripts presenting original results or novel treatments of existing results are particularly encouraged. All volumes published in SBASE undergo a thorough refereeing process.

The SBASE series is published under the auspices of the German Statistical Society (Deutsche Statistische Gesellschaft).

Norbert Hirschauer • Sven Grüner •
Oliver Mußhoff

Fundamentals of Statistical Inference

What is the Meaning of Random Error?

 Springer

Norbert Hirschauer (iD)
Institute for Agricultural and
Nutritional Sciences
Martin Luther University
Halle-Wittenberg
Halle (Saale), Germany

Sven Grüner (iD)
Institute for Agricultural and
Nutritional Sciences
Martin Luther University
Halle-Wittenberg
Halle (Saale), Germany

Oliver Mußhoff (iD)
Department for Agricultural and
Rural Development
Georg-August-University Göttingen
Göttingen, Germany

ISSN 2524-4116   ISSN 2524-4124   (electronic)
SpringerBriefs in Applied Statistics and Econometrics
ISBN 978-3-030-99090-9   ISBN 978-3-030-99091-6   (eBook)
https://doi.org/10.1007/978-3-030-99091-6

This Springer imprint is published by the registered company Springer Nature Switzerland AG
The registered company address is: Gewerbestrasse 11, 6330 Cham, Switzerland

[...] in most scientific settings, the arbitrary classification of results into 'significant' and 'non-significant' is unnecessary for and often damaging to valid interpretation of data; [...] estimation of the size of effects and the uncertainty surrounding our estimates will be far more important for scientific inference and sound judgment than any such classification.

Greenland et al. (2016: 338): Statistical tests, P values, confidence intervals, and power: a guide to misinterpretations

Preface

In modern life, data-based scientific propositions about the world are both omnipresent and extremely important for sound decision-making in organizations and society as a whole. Think of climate change or the COVID-19 pandemic with questions such as of how face masks, vaccines, or regulatory restrictions on people's movements work. Properly done studies help us accumulate knowledge and form the most reasonable propositions about (unknown) real-world facts—under due consideration of the inevitably remaining uncertainty. In contrast, poorly done studies as well as misinterpreted study results obstruct sound judgments and may instigate overconfident conclusions that lead us astray.

Statistics can provide valuable tools for inferring something meaningful about a broader context based only on a limited dataset, such as a random sample drawn from a larger population. However, the conventional routine of statistical significance testing seems to cause confusion rather than help researchers make sound judgments on a study's knowledge contribution under consideration of the uncertainty resulting from random error. Despite long-standing criticisms, misapplications and misinterpretations of statistical significance tests continue to be an alarmingly "normal" practice. The problems seem to be here to stay even though the scientific debate on these issues has intensified over the last few years—as highlighted, for example, by the unprecedented methodological warning of the *American Statistical Association* in 2016 and the call to retire statistical significance in *Nature* in 2019.

This book is motivated by a deep concern about the uninterrupted stream of misleading significance declarations and overconfident proclamations of scientific discoveries (dichotomous yes/no conclusions) that damage the reputation of science and cause the risk of ill-informed choices. But this book does not concern itself with those inferential errors from the outset. Instead, before discussing the misinterpretations that are so intimately associated with dichotomous significance declarations, we go back to the conceptual foundations of statistical inference. In other words, we show that what we have, at best, is an unbiased data-based estimation of an effect size (e.g., a gender pay gap or a relationship between education and income in a population) and an equally data-based estimation of the "uncertainty surrounding our estimates" caused by random error.

We have made a number of contributions to the methodological debate dealing with the question of how to preclude inferential errors resulting from significance testing routines (some say, "rituals") that are performed without a second thought and consideration of the specific research context. Each of these contributions covered different aspects of the problem. Leaning heavily on these materials, this book develops a coherent description of the conceptual foundations of statistical inference and shows how changes for the better could be brought about. In particular, we used material from the following papers, for which we either retain ownership of the copyright under the Creative Commons Attribution License or for which permission has been granted by the copyright holders:

- Hirschauer, N. (2022): Unanswered questions in the p-value debate. Significance (June 2022): 42–44 (https://doi.org/10.1111/1740-9713.01655).
- Hirschauer, N., Grüner, S., Mußhoff, O., Becker, C., Jantsch, A. (2021b): Inference using non-random samples? Stop right there! Significance (October 2021): 20–24 (https://doi.org/10.1111/1740-9713.01568).
- Hirschauer, N., Grüner, S., Mußhoff, O., Becker, C. (2021a): A Primer on p-Value Thresholds and α-Levels – Two Different Kettles of Fish. German Journal of Agricultural Economics 70: 123–133 (https://doi.org/10.30430/70.2021.2.123-133).
- Hirschauer, N., Grüner, S., Mußhoff, O., Becker, C., Jantsch, A. (2020b): Can p-values be meaningfully interpreted without random sampling? Statistics Surveys 14(2020): 71–91 (https://doi.org/10.1214/20-SS129).
- Hirschauer, N., Grüner, S., Mußhoff, O., Becker, C. (2020a): Inference in economic experiments. Economics. The Open-Access, Open-Assessment E-Journal 14(2020-7): 1–14 (https://doi.org/10.5018/economics-ejournal.ja.2020-7).
- Hirschauer, N., Grüner, S., Mußhoff, O., Becker, C. (2019): Twenty steps towards an adequate inferential interpretation of p-values in econometrics. Journal of Economics and Statistics 239(4): 703–721 (https://doi.org/10.1515/jbnst-2018-0069).
- Hirschauer, N., Grüner, S., Mußhoff, O., Becker, C. (2018): Pitfalls of significance testing and p-value variability: An econometrics perspective. Statistics Surveys 12(2018): 136–172 (https://doi.org/10.1214/18-SS122).
- Hirschauer, N., Mußhoff, O., Gruener, S., Frey, U., Theesfeld, I., Wagner, P. (2016): Inferential misconceptions and replication crisis. Journal of Epidemiology, Biostatistics, and Public Health 13(4): e12066-1–e12066-16 (https://riviste.unimi.it/index.php/ebph/article/view/17632).

Many people supported us in writing this book, with their reading, critical questions, and suggestions. Above all, they challenged us to have students and statistical practitioners with various statistical backgrounds in mind. We are especially grateful to our colleagues at the Martin Luther University Halle-Wittenberg and the Leibniz Institute of Agricultural Development in Transition Economies in

Halle who asked many helpful questions and provided encouragement during the meetings of our informal working group on statistical inference. In particular, we would like to thank the coauthors of our inference-related papers who generously allowed us to use them as a quarry for this book: Claudia Becker, Ulrich Frey, Antje Jantsch, Insa Theesfeld, and Peter Wagner.

Halle (Saale), Germany   Norbert Hirschauer
Halle (Saale), Germany   Sven Grüner
Göttingen, Germany   Oliver Mußhoff

Contents

Abbreviations

AEA	American Economic Association
ASA	American Statistical Association
CI	Confidence interval
fpc	Finite population correction (factor)
H_0	Null hypothesis
H_A	Alternative hypothesis
HARKing	Hypothesizing after results are known
MAR	(Data) missing at random
MCAR	(Data) missing completely at random
MNAR	(Data) missing not at random
NHST	Null-hypothesis-significance-testing
PATE	Population average treatment effect
PSU	Primary sampling unit
RCT	Randomized controlled trial
RDD	Regression discontinuity design
SATE	Sample average treatment effect
SE	Standard error
SRS	Simple random sample or simple random sampling (context dependent)
SSU	Secondary sampling unit
WEIRD	Western, educated, industrialized, rich, and democratic

Chapter 1
Introduction

Data on how many scientific findings are reproducible are bleak, and in recent years a wealth of papers have warned against "false findings" that jeopardize the reputation of science and cause harmful effects for society. The reasons for false discoveries are manifold, but misuses and misinterpretations of p-values and statistical significance tests are the most prominently decried ones. While substantial criticisms have been expressed for more than half a century, inferential errors associated with the widely entrenched convention of null-hypothesis-significance-testing are ubiquitous. Basing narrative reviews on "vote counting" is one of the many devastating but widespread fallacies. Such reviews contrast the tallies of "significant" ("positive") and "non-significant" ("negative") findings to assess whether research in a certain field rather confirms or contradicts the existence of an effect. It is an ethical imperative and important part of good scientific practice to try to avoid such fallacies and ensuing harms as best as we can. Attempting an apt judgment of what one should most reasonably believe in the light of the available evidence and the inevitably remaining uncertainty is an extremely important task in any field of science.

The Deplorable State of Statistical Inferential Practice
In the light of persistent and prevalent misunderstandings associated with p-values, the *American Statistical Association* felt compelled to issue an unprecedented methodological warning in 2016. It stressed that the p-value and statistical significance testing can neither be used to determine whether a hypothesis is true nor whether a finding is important (Wasserstein & Lazar, 2016). Three years later, *The American Statistician* published a special issue "Statistical Inference in the 21st Century: A World Beyond $p < 0.05$." Summing up the reform suggestions, the editors state that it is time to abandon statistical significance testing (Wasserstein et al., 2019). Almost simultaneously, *Nature* published a widely supported call to "Retire statistical significance" (Amrhein et al., 2019). In the same year, the National Academies of Sciences (2019) took up criticisms regarding inferential errors in its "Consensus Report on Reproducibility and Replicability in Science."

N. Hirschauer et al., *Fundamentals of Statistical Inference*, SpringerBriefs in Applied Statistics and Econometrics, https://doi.org/10.1007/978-3-030-99091-6_1

Despite prominent institutional efforts for reforms, null-hypothesis-significance-testing remains the predominant statistical routine that is often ritualistically performed without a second thought and without consideration of the specific research context and data. As a consequence, we face an uninterrupted stream of misleading significance declarations and overconfident proclamations of scientific discoveries (yes/no conclusions). This alarmingly "normal" practice is largely due to two superordinate aberrations: first, "neat" and easy-to-swallow misinterpretations have been perpetuated for decades through inappropriate teaching and even best-selling statistics textbooks under the disguise of mathematical exactitude. Second, even leading journals that are widely considered as beacons for best practice enforce compliance with the flawed ritual and accept or actually call for bold proclamations of novel and "significant discoveries." Inappropriate teaching and flawed journal policies have effectively propagated significance testing as *the* inferential standard, even though it causes egregious inferential errors and obstructs critical reasoning.

The Book's Aims and Intended Audience
This book is based on the premise that a lacking understanding and internalization of the probabilistic foundations of statistical inference is responsible for the ritual-like adherence to significance testing procedures and associated misinterpretations. Therefore, we go back to the roots and aim at providing readers with a working intuition and conceptual understanding of statistical inference. We hope that an intuitive conceptual grasp will reduce researchers' susceptibility to ill-understood routines and misinterpretations (cognitive biases). Cognitive biases are virtually invited by the delusive language of conventional statistics which often contradicts natural language meanings. Confusing the term "statistically significant" with "large" or "scientifically proven" is only one example. Unfortunately, the vast but scattered body of literature that has dealt with *p*-value-related misuses and mis-interpretations over the last six decades has been of little avail so far. A coherent description of the conceptual foundations of statistical inference that leaves behind ill-founded statistical conventions is, therefore, essential for substantial changes for the better. Setting aside sophisticated approaches such as Bayesian statistics or specific methods (e.g., for big data), this book provides such a coherent descrip-tion—with a particular focus on frequentist statistical practice in the social sciences.

The representations throughout the book are as non-technical as possible. This strategy is to provide a broad readership with a clear understanding of what statistical inference presupposes, what it can do, and what it cannot do. The book is generally intended for readers with an interest in understanding the role of statistical inference in a prudent assessment of the knowledge gain that can be obtained from a particular set of data. More specifically, the book provides an accessible resource for graduate students as well as statistical practitioners who have a basic knowledge of statistics. The first parts of the book that deal with validity and the probabilistic foundations of statistical inference—random sampling error, sampling distribution, estimation uncertainty—could also be used in an introductory statistics course. But, above all, the book is a resource for statistical practitioners who are confronted with the methodological debate about the drawbacks of "significance testing" but do not know what to do instead. It is also targeted at scientists who have a genuine methodological interest in the reform debate regarding "A World Beyond $p < 0.05$." Both practitioners and scientists with a methodological bent will profit from the systematic description of what statistical inference presupposes and of what it can accomplish when the information that can be derived from a randomly generated dataset is not downgraded into dichotomous significance declarations.

Being didactically focused on the fundamental question of what we can reasonably infer about a population based only on a random sample of that population, we avoid specific detail as much as possible. Therefore, we elaborate neither on complex data structures nor questions of advanced model specification. For example, we focus on randomly obtained cross-sectional data to facilitate the understanding of the basic probabilistic principles of statistical inference as such. However, these principles also apply to other data structures such as longitudinal data (e.g., panel data) as long as they are randomly generated and thus allow assessing the uncertainty of an estimation caused by random error. Similarly, we do not deal with the question of which mathematical manipulation of the data (model specification) adequately describes a data pattern, such as relationships between variables, in the first place. Consequently, we only slightly touch upon even the simplest regression model (simple linear regression); and we do so only to illustrate that *all* statistics, including regression slopes, are first of all summary descriptions of the given dataset. While the determination of a sample regression slope as such is commonly known as "estimation," we avoid this delusive terminological usage. A sample slope is first of all a description of a X-Y-relationship in the given dataset. This *sample* statistic can only be used as an *estimate* for an unknown *population* relationship when there is a *probabilistic* data collection design.

It should also be observed that we do not aim at providing an encyclopedic overview of the many inferential formulas needed in various research contexts. Instead, we are highly selective in what we present, following our strategy to provide a parsimonious and accessible introduction to foundational matters of statistical inference. This has several implications for the book's content: first, we focus on simple random sampling and observational study throughout most of the book, but we separately address more complex sampling designs, such as stratified and cluster sampling (Chap. 4), as well as randomized controlled trials (Chap. 7). Second, we use formal mathematical representations only to support the conceptual comprehension of the potential and limitations of statistical inference. For example, we display the relatively simple formulas for the standard error of the mean for different sampling designs to convey the basic fact that the sampling design influences the uncertainty of an estimation. But we refrain from displaying the more complex standard error formulas for other estimates (e.g., mean differences, regression slopes) for sampling designs other than simple random sampling.

Outline of the Book's Contents

The book contains eight chapters. After this introduction, **Chap. 2** outlines the concept of validity and describes the relationship between statistical and scientific inference. Scientific inference can be broadly defined as the adequate assessment of the informative value ("validity") of an empirical finding. We show that statistical inference is only part of the much larger enterprise of scientific inference. This is because, among the many errors that restrict and jeopardize the validity of empirical findings (e.g., random sampling error, non-random sampling error, measurement error, specification error), statistical inference only deals with random error.

Chapter 3 describes the probabilistic basics of statistical inference. Using the case of simple random sampling, we show that the standard error is but an estimate of the standard deviation of the sampling distribution, i.e., the hypothetical distribution of all sample-based estimates that we would find over frequently repeated random draws of equal-sized samples (hypothetical statistical replications). We show that statistical inference strictly presupposes a probabilistic process of data generation (random sampling) and considers random error when making inferences from the idiosyncrasies of the particular sample to its parent population. The chapter's key message is that what we can extract from a particular set of data is, at best, an unbiased point estimate (a meaningful "signal") and an unbiased estimate of the uncertainty of this point estimation ("noise") quantified by the standard error.

Chapter 4 elaborates on the fact that the estimation of effect sizes and standard errors requires considering the specific sampling design, such as simple random sampling, stratified sampling or cluster sampling. In other words, we need to consider the probabilistic mechanism through which members of the population are selected into the sample. Chapter 4 also shows that convenience samples, which are often tacitly treated as if they were random samples, violate the probabilistic precondition of statistical inference: we ignore and can, therefore, not consider the selection mechanism. The resulting sample selection bias precludes statistical inferences toward a broader population because we do not know to what extent sample members are systematically different from other members of that population. Given the widespread use of convenience samples in the social sciences, we give a basic introduction to sample selection models that are aimed at correcting for selection bias in non-random samples. We also scrutinize the meaning of statistical inference in cases where we already have data from entire populations.

Chapter 5 emphasizes that, due to the law of large numbers, the uncertainty of an estimation resulting from random sampling error decreases the larger the size of the sample or the larger the number of samples that we consider. We might also say that the inferential leap from the data to the population decreases when we have more observations. Accounting for this simple fact, we give a basic introduction to meta-analysis. We show how meta-analysis can be used to synthesize the point and standard error estimates of single studies into summary statistics ("meta-statistics") that provide the most appropriate picture of the real-world state of interest. Meta-analysis exploits the fact that unbiased estimators estimate correctly on average over frequently repeated draws of random samples. Therefore, it doesn't make sense to ask whether an individual study's estimate is "true" or not. Instead, we need to consider the body of evidence and include the appropriately weighted knowledge contribution of each properly made study even if, by its own, it only produces an estimate with a large uncertainty.

The book's most extensive **Chap. 6** relates the probabilistic basics of statistical inference to the methodological debate about p-values and statistical significance. We first describe the p-value and the null-hypothesis-significance-testing (NHST) approach and then identify their drawbacks and pitfalls. NHST downgrades the two meaningful pieces of information that we can extract from a random sample—the point estimate (signal) and the uncertainty of the estimation (noise)—first into a quotient (signal-to-noise ratio), then into a p-value (based on the usually uninformative null hypothesis of zero effect), and finally into a dichotomous significance declaration (based on an arbitrary p-value threshold such as 0.05). We show that nothing is gained by this downgrade. On the contrary. The associated jargon that speaks of "significant" ("positive") as opposed to "non-significant" ("negative") results is delusive and makes not only the consumers of research but also many researchers draw rash yes/no conclusions from individual studies.

Given NHST's poor track record, we also dive back into history to explain how it came that the meaningful signal and noise information that can be extracted from a random sample was distorted almost beyond recognition into statistical significance declarations. It seems that a "wrong historical bend" in the wake of semantic confusions has led to an amalgamation of two irreconcilable approaches: "significance testing" by Fisher (1925) and "hypothesis testing" by Neyman and Pearson (1933a, 1933b). Getting acquainted with the two original perspectives elucidates why cobbling them together causes inferential errors and why *moving forward* in statistical inference requires *going back to the roots* and using the standard error as a measure of the uncertainty of sample-based estimates.

Going beyond random sampling and generalizations from samples toward populations, **Chap. 7** describes the role of statistical inference in randomized controlled trials. Randomized experimental designs are based on a deliberate intervention (treatment) and a random assignment of experimental subjects to treatments. Randomization balances known *and* unknown confounders across treatment groups and thus ensures that everything but the treatment is held constant (ceteris paribus condition). Ensuring ceteris paribus facilitates using an observed mean difference as an unbiased estimate of the causal treatment effect. Being a probabilistic process, randomization enables researchers to use the standard error to assess the uncertainty of the treatment effect estimation. But unlike random sampling, the standard error now describes the estimated standard deviation of the randomization distribution, i.e., the distribution of all mean differences that we would see over frequently repeated re-randomizations (random re-groupings) of the experimental subjects.

Finally, **Chap. 8** discusses how statistical inference should look like in the future in order to fully exploit its potential while strictly avoiding "neat" but wrong conclusions that are so intimately invited by the statistical ritual NHST.

What the Reader should Consider

Having warned against semantically induced misunderstandings, a few comments regarding the book's **terminology** must be made. First of all, it is worth observing that we use the term "effect size" for the sake of brevity. No causal meaning is implied. Similarly, the pair of terms "independent/dependent variable" and related synonyms (explanatory/response, input/output, exogenous/endogenous, etc.) are not meant to imply a cause-effect relationship, unless explicitly stated otherwise.

The term "random sampling" could also be a source of misunderstanding. Sometimes, this label is already attached to researchers' intention to sample at random by using lottery-like selection procedures. However, such procedures rarely guarantee that we obtain random samples. For example, in voluntary surveys, researchers cannot force randomly selected subjects to participate. When the propensity to respond differs between individuals, which is often the case, we face the problem of self-selection. As a result, members of the population are not randomly selected into the sample despite researchers' lottery-like selection procedures. That is, sample members and non-members might be systematically different because the overall resulting process of data generation is not probabilistic. Since we focus on statistical inference, which presupposes a random process of data generation, we use the term "random sampling" only when we succeed in *obtaining* a random sample.

Throughout the book, we avoid cumbersome statistical jargons as much as possible. Our terminology is particularly aimed at supporting the conceptual understanding that statistical inference is based on probability theory and concerned with the question of what we can learn from a quantity observed in a particular random sample (i.e., a sample-based "estimate") about the corresponding population quantity—under due consideration of the estimation uncertainty caused by random error. Compared to conventional terminology, we adopt a more restricted use of the word "estimation" and related terms. We only use these terms when data collection was demonstrably probabilistic (e.g., in the case of random sampling). In all other cases, we simply speak of summary statistics that describe certain properties of the data at hand. For example, we keep to calling the arithmetic mean of observed data a "summary descriptive statistic" or simply a "sample quantity." But we do *not* refer to this quantity as "estimate" of the population mean unless the sample was randomly drawn from that population. The same applies to other quantities such as a mean difference or a between-variable relationship such as a regression slope. Along those lines, we do not attach the label "estimator" to the mathematical rule for computing these sample quantities unless the underlying data were randomly generated.

This restrictive terminological usage is to increase the awareness that we must not rashly lurch from the description of some conveniently available data to overconfident generalizations beyond the confines of a particular, and potentially non-random, sample. For illustration sake, imagine we conveniently collected a non-random sample of individuals at a gym that happens to comprise many basketball players. While we can, of course, compute the quantity "mean body height" for the observed data, we would, in this case, neither call the mathematical rule of computing the mean an "estimator" nor the obtained result an "estimate" because we are not dealing with a random-sample-based statistic. In brief, we equate "estimation" with "unbiased statistical estimation" based on random data, which is the precondition for a meaningful use of inferential statistical procedures. When we have reason to suspect that we can infer little about a population because we have a sample that is systematically unrepresentative of that population, we either avoid the word "estimate" (and related terms) altogether or clearly qualify the respective sample statistic as an inadequate or *biased* estimate. That said, it should be observed that we shortly speak of "biased/unbiased estimates" instead of using the lengthy formulation "estimates resulting from a biased/unbiased estimation."

The conceptually motivated terminological usage described above is reflected in the **notation**, which underpins the distinction between descriptive sample quantities, estimates, and population quantities: lowercase Latin letters denote descriptive sample quantities. We stick to this notation as long as we have not argued and proved that there was a probabilistic data collection design. For example, $\bar{x}$ denotes the arithmetic mean of the observations in a particular sample—irrespective of whether we deal with a random sample or not. Similarly, in regression analysis, b denotes the association between two variables in the sample ("sample regression slope"). *If*, and only if, descriptive quantities such as the sample mean $\bar{x}$ and the sample regression slope b are derived from a random sample, we interpret them as (statistical) estimates of the unknown population quantities, in which case we write: $\bar{x} = \hat{\mu}$ and $b = \hat{\beta}$. More generally speaking, we use lower case Greek letters accompanied by "hat" (e.g., $\hat{\mu}$, $\hat{\beta}$) to denote random-sample-based statistics that can indeed be used as estimates. Finally, simple lower case Greek letters denote the unknown population quantities of interest. For example, μ denotes a variable's unknown population mean, and β denotes the unknown relationship between two variables in the population ("population regression slope") that we are interested in estimating based on sample observations.

As regards notation, it should also be observed that we do not formally distinguish between "estimate" and "estimator." The meaning will be clear in the respective context. The term "estimator" can generally be understood as the variable name that is attached to a random-sample-based statistic (e.g., a mean, a mean difference, or a regression slope) for which we will obtain different values across different random samples. In contrast, the term "estimate" denotes the numerical value of this statistic that we observe in one particular random sample. For ease of notation, we refrain from using subscripts or superscripts to distinguish a particular sample-based estimate from the estimator (the variable). For example, when looking at a particular random sample, $\hat{\beta}$ denotes the association between two variables as estimated from the observed data. But in another context, for example when we look at the sampling distribution of all estimates that we would obtain over frequently repeated random draws of equal-sized samples, $\hat{\beta}$ denotes the estimator. Generally, we use detailed notation when necessary but go "shorthand" whenever there are no ambiguities. For example, $\widehat{SE}$ is used for the estimated standard error of a regression slope when it is clear that we talk about a regression slope. But we distinguish the estimated standard error of a regression slope $\widehat{SE}_b$ from the estimated standard error of a mean difference $\widehat{SE}_d$ when necessary to avoid confusion.

Finally, all main chapters conclude with a **"Chapter summary"** and a **"Recommended reading"** list that is to guide further study. Readers are warned, however, that many otherwise valuable textbooks and papers still cobble the approaches by Fisher and Neyman-Pearson together and heavily focus on NHST. Critical reading is necessary to forestall falling prey to the erroneous yes/no conclusions that are so intimately associated with significance testing procedures. Beyond the "Recommended reading" list, references are provided where appropriate. Depending on content, the intensity of referencing varies from section to section. For example, we refrain from providing references when describing simple statistics such as the mean or a mean difference. But we provide detailed references for more specific methodological issues. Sample selection models and meta-analyses are examples.

Chapter 2
The Meaning of Scientific and Statistical Inference

2.1 The Starting Point: Errors and the Assessment of Validity

In empirical research, **scientific inference** can be defined as an act of reasoning aimed at arriving at the most reasonable belief (opinion, judgment, proposition) regarding a real-world state of interest—in the light of all available evidence. A more dynamic perspective would emphasize that the scientific enterprise itself is about continuously accumulating evidence and knowledge. Consequently, scientific inference can be understood as the *process* of moving from a prior scientific belief, which was the most reasonable one before a new piece of information turned up, to another scientific belief, which becomes reasonable after due consideration of this new piece of information.

Moving to "another scientific belief" can simply mean that the uncertainty regarding the real-world state or relationship of interest (e.g., the gender pay gap or the education–income relationship in a certain country) is reduced because a new study produced findings that are consistent with previous ones. But it can also mean that a prior scientific belief is questioned or must be updated as regards content because a new study produced an outcome that substantially differs from previous ones. Incremental information is not only provided by studies that collect and analyze new (primary) data but also by studies that extract new information from existing (secondary) data.

A definition by Ramsey and Schafer (2013: 8) describes scientific inference in quantitative empirical research as "a conclusion that patterns in the data are present in some broader context." This definition might cause confusion, at least when, in its brevity, it is taken to convey a dichotomous perspective. Understanding inference as a yes/no conclusion that patterns observed in a particular dataset are *either* present *or not* in a broader real-life context does not reflect its true nature. Inference is an act of reasoning that explicitly considers the **inevitably remaining uncertainty** regarding the real-world state of interest. However, the definition's brevity also has an

N. Hirschauer et al., *Fundamentals of Statistical Inference*, SpringerBriefs in Applied Statistics and Econometrics, https://doi.org/10.1007/978-3-030-99091-6_2

advantage. It pointedly brings out the fact that researchers are rarely interested in the patterns (or, "effect sizes") that can be found in a particular set of data. Instead, identifying effect sizes in a set of data is usually but a means to an inferential end.

Although the end of a data analysis is nearly always to learn as much as possible about a broader **inferential target context**, "as much as possible" will regularly be much less than what is needed for a conclusive, final judgment. Imagine a survey in which 100 farmers who visit a local agricultural fair are interviewed regarding their income. Unless one is really interested in exactly and only these 100 haphazardly interviewed farmers, one would first have to define the study's end. Is it aimed at understanding the income situation of farmers in the district or in the federal state where the fair was held? Or is it aimed at a still broader population such as the farmers in the whole country? Once the end is defined, the critical inferential question arises of what one can reasonably—i.e., under due consideration of the uncertainty from various sources—claim to learn from the patterns among the 100 interviewed farmers for the broader target context. Besides mean income, exemplary patterns of interest might be differing incomes between male and female farmers, or the relationship between education and income, or patterns for other variables such as the differing willingness to participate in environmental programs among various subgroups of farmers.

Scientific inference can also be briefly defined as the rational assessment of the informative value of an empirical finding. This informative value, which is also called the **validity** of a finding, is restricted and jeopardized by various types of random and non-random **errors**, such as random sampling error, non-random sampling error, measurement error, and specification error.

The distinction between different error types can be used to describe the relationship between scientific inference and statistical inference: **statistical inference** is only a small part of the much larger enterprise of scientific inference. This is because, among the many errors that can arise in empirical research, statistical inference can only deal with random error, which, in observational studies, exists only when there is random sampling. Statistical inference is based on probability theory; to be more precise, statistical inference is based on a formal probability model that links a randomly generated dataset to the broader target population and helps us evaluate the single study's knowledge contribution in the light of the estimation uncertainty ("noise") resulting from **random error**.

Figure 2.1 provides a systematization of validity on different hierarchical levels. The figure illustrates which types of error limit a study's knowledge contribution and jeopardize which types of validity. The most basic classification is to differentiate between **external validity** (i.e., the informative value of sample observations for a broader population and context) and **internal validity** (i.e., the informative value of sample observations for the sample and study context itself). External validity, which is the core issue of this book, will be discussed in detail in Sect. 2.2. Internal validity will be covered in Sect. 2.3.

Figure 2.1 also includes randomization error as a source of uncertainty that limits the validity of a data-based estimation. Randomization error impairs causal (internal) validity in randomized controlled trials (RCTs), which are aimed at estimating

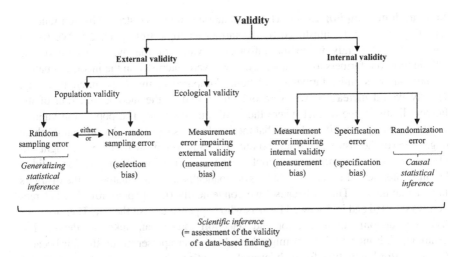

Fig. 2.1 The concept of validity

average causal treatment effects within a given group of experimental subjects. While we treat causal inference in RCTs largely as a separate issue in Chap. 7, we included randomization error in the comprehensive systematization of the various types of validity and error in Fig. 2.1. This is to bring out the crucial fact that random errors (i.e., **random sampling error** and **randomization error**) are the only sources of uncertainty that inferential statistical procedures, being based on probability theory, can deal with.

2.2 External Validity

External validity describes the informative value that a finding in a particular dataset has for some broader context. Gerber et al. (2004) note that external validity generally increases the smaller the "inferential leap" from the idiosyncrasies of a particular study setting and dataset to the real-life context of interest. Within external validity, two subordinate validity dimensions can be distinguished (cf. Fig. 2.1): **population validity** describes the informative value that a data-based finding has for the real-life population of interest. Simply put, population validity is concerned with generalizations from samples to populations. **Ecological validity** describes the informative value that a data-based finding has for the real-life setting of interest. Simply put, ecological validity is concerned with generalizations from more or less artificial study settings toward real-life settings. Within each of these two dimensions, different errors compromise validity: the larger the inferential leap, the larger these errors; and the larger the errors, the lower the validity of a pattern identified in a particular dataset for the broader inferential target context.

Non-random Sampling Error How population validity is restricted by non-random sampling error can be illustrated by the farmer example from Sect. 2.1. We would intuitively and rightly judge the informative value of the mean income of the 100 interviewed farmers to be rather low if we were interested in the income situation of the country's entire farmer population. Compared to this large inferential leap, validity would increase if we were only interested in the income situation of the farmers living in the district where the local fair was held. The population validity would increase further if we knew that the 100 farmers represented a substantial share (e.g., 50%) of the district's farming community. But even in such a case, we might still observe a mean income in the sample that has a questionable population validity because interviewed farmers might be systematically unrepresentative of the district's farmer population. This is because, we conveniently (i.e., haphazardly but not randomly) selected and interviewed farmers who happened to visit the fair. Therefore, we face non-random sampling error and **selection bias** to an unknown degree. For example, high-income farmers might have been overrepresented on the fair because they can afford more time for such leisurely activities. The income in this convenience sample of haphazardly selected farmers is likely to provide a distorted view of the income in the target population, i.e., convey little information regarding the income situation of the district's farmer population. Unfortunately, we cannot know if and to which degree selection bias is present without having information beyond the convenience sample.

Random Sampling Error Population validity would considerably increase if interviewed farmers represented a random sample from the district's farmer population. For example, a lottery-like procedure giving each farmer in the district equal probability of being selected into the sample, in conjunction with 100% survey participation of those who are selected, would produce a random sample. The mean farm income in such a sample could be used as an unbiased estimate of the mean farm income in the district. While selection bias would be eliminated in the case of random sampling, the estimation of population quantities (here, the mean income in the district) would still be uncertain due to random sampling error, i.e., the fact that what we see in a random sample does not exactly reflect the properties of the population. In other words, sample quantities deviate more or less from the corresponding population quantities because of random sampling error. **Generalizing statistical inference** denotes the statistical procedures that are used to assess the **estimation uncertainty** (noise) resulting from random sampling error. Due to the law of large numbers, random sampling error decreases when sample size increases. We might also say that the inferential leap from data to a population decreases when there are more observations. Random sampling error completely disappears when we can study full populations. Consequently, there is neither need nor room for generalizing statistical inference when we have data for an entire population.

Measurement Error Impairing External Validity Ecological validity, the second dimension of external validity, tackles the question of what we can learn from observations in an artificial research setting for the real-life setting of interest. Again, the farmer example provides intuition. Rather than, say, using anonymized farm income data from a source such as the tax authority, data were collected through interviews at a local fair. For one reason or the other, farmers' statements in this setting might not reveal their true income. Ecological validity would increase, for example, the more confident we could be that farmers had no reason to conceal their income. This assumption is not trivial. Farmers might have systematically understated their income because they expected the study to influence political decisions regarding future transfer payments to the farming community—with the hope of receiving higher payments when the study finds farmers' income situation to be bleak. More generally said, data obtained in a particular research setting with a particular measurement instrument might be systematically distorted and not reveal the real-world state of interest even if we have a random sample. This can also be understood as **measurement bias** impairing external (ecological) validity. When farmers' answering behavior in the interview setting conveys little information regarding their factual income (low ecological validity), it makes little sense to ask the question of whether we can generalize toward a larger population. Using inferential statistics in such a situation would imply concerning oneself with the inaccurate answers that the population would give in the artificial research setting.

We may summarize that high external validity (i.e., a high informative value of a pattern in a particular set of data for a broader target context) is achieved when high population validity (small sampling error) as well as high ecological validity (small measurement error) can be ensured. In the assessment of external validity, there can be a role for statistical inference. However, this role is limited because neither measurement error nor non-random sampling error can be assessed through inferential statistics. Statistical inference can only deal with random error. Therefore, it can only help us evaluate the external validity of a sample-based estimation *if* the data are generated through random sampling *and if* ecological validity can be assumed to be satisfactorily ensured.

2.3   Internal Validity

Rather than looking toward a broader context, internal validity, the second basic dimension of validity (cf. Fig. 2.1), stays within the confines of the given study setting and sample. In brief, one could describe internal validity as being concerned with the question of what the observations in a sample can teach us about the sample in the study context itself. Analogous to external validity, different types of error compromise internal validity; and the larger these errors, the lower internal validity.

Measurement Error Impairing Internal Validity   Similar to external (ecological) validity, internal validity can be impaired by measurement error. At first view, the distinction between measurement error impairing external validity as opposed to measurement error impairing internal validity may not be obvious. Compared to observational studies, the distinction is indeed more salient in experimental studies where external (ecological) validity deals with the straightforward question of whether the behaviors observed ("measured") in an artificial experimental setting such as a lab are a good indication of subjects' real-life behaviors. But we can also illustrate the difference in the farmer example. A **measurement bias** impairing internal validity arises if the measurement instrument itself is inappropriate. The device to measure farmers' income was the particular wording used by a particular interviewer in an oral interview. On the one hand, there might be mishearing and/or data entry error on the part of the interviewer. On the other, farmers might have ill-understood the wording and interpreted it in different ways. For example, some may have provided a figure for the average per capita income of all members of the farming family while others may have provided the total income of the farming family. If so, the data are quite worthless even when there are no ecological validity issues putting into question whether farmers' answering behaviors in the specific setting of the interview revealed their true income. It is important to note that a correct ("accurate") measurement within the given research context is the indispensable precondition for meaningful further-reaching inferences—whether they are toward a broader real-life setting or toward a broader population or both.

Specification Error When the outcome of interest is an association between two variables, a special threat to internal validity arises from specification error. A specification error occurs, for example, when one or several (confounding) variables other than X modify the outcome of interest Y but are not considered ("controlled for") in the regression model used for identifying the X-Y-relationship. An ill-specified model with lacking control of confounders leads to a distorted (upward- or downward-biased) X-Y-association (**specification bias**). Similar to measurement error, specification error must be excluded before thinking beyond the given dataset. Imagine we are interested in the relationship between education X and income Y among the 100 interviewed farmers. Imagine also that, besides education, farmers' income is affected by factors such as soil quality or professional experience. Given such patterns in the data, using a simple linear regression model with only education X as explanatory variable is likely to distort the education-income relationship. It would be quite meaningless to think about external validity and generalizations to a broader context when such distortions are likely to be present.

Randomization Error In accordance with the book's primary focus, we have so far discussed the meaning of inference and validity in observational studies. An internal validity issue in its own right arises in experimental studies where participants are randomly assigned to a control (non-treatment) and a treatment group (RCTs). The aim of random assignment ("randomization") is to generate equivalence between groups and thus facilitate the estimation of an average causal treatment effect in a ceteris paribus setting. However, only "*statistical* independence of treatments" is ensured by the probabilistic process of randomization. That is, the estimation of a causal treatment effect remains uncertain due to randomization error. The term "**causal statistical inference**" denotes the statistical procedures that are used to assess the **estimation uncertainty** regarding a causal effect resulting from randomization error. Imagine we randomly assigned the 100 haphazardly selected farmers to two equal-sized groups. Imagine also that the control group was directly asked the income question, whereas the treatment group was first told that the study's objective was to identify farmers' need for transfer payments. Next, assume we find the mean stated income in the control group to be 500 above the mean in the treatment group. This difference would be an unbiased estimate of the average causal effect that the treatment "information about the study's objective" has on the answering behavior of the given sample of 100 farmers ("sample average treatment effect"). However, due to the vagaries of randomization, more or less farmers with higher incomes may be assigned to one or the other group in any particular random assignment. In different random re-groupings of the 100 farmers, we would have found different mean differences even though there is only one true causal effect in the given group. Due to the law of large numbers, the uncertainty of a randomization-based causal effect estimation decreases when sample size increases. For intuition sake, imagine a much larger sample of 10,000 farmers were randomly assigned to a control and a treatment group. In this case, the vagaries of randomization would produce less severe variations across hypothetical re-randomizations and, therefore, leave us with less uncertainty regarding the average treatment effect.

2.4   Chapter Summary: Scientific Inference Is More Than Statistical Inference

A final comprehensive look at Fig. 2.1 highlights the essentials of the multi-dimensional validity problem and carves out the role of statistical inference within the larger enterprise of scientific inference:

1. Scientific inference is about accumulating knowledge by drawing reasonable conclusions regarding a real-world state of interest given prior evidence *and* the incremental information that was extracted from the dataset under study. In brief, scientific inference can be said to be concerned with the validity (informational value) of a data-based finding for a relevant broader context.
2. Statistical inference is only part of the much larger enterprise of scientific inference. This is because, among the many errors that threaten validity in empirical research, statistical inference deals only with random error.
3. Statistical inferential quantities can only be used to assess how random error affects the validity of an empirical finding if there was a random process of data generation—either random sampling or randomization or both.
4. In the larger enterprise of scientific inference, statistical inference has the role of helping evaluate the single study's knowledge contribution under consideration of the noise from random error, i.e., by assessing the estimation uncertainty caused by random error.
5. In observational studies, random sampling error would completely disappear if we could study the full population. In contrast, non-random errors can cause serious validity problems even when we are in the comfortable position of having access to the full population.

6. Even if there is a random process of data generation, random error is often not the most important error compromising a study's validity. Non-random errors such as measurement and model specification errors are often far more problematic issues.
7. The meaning of statistical inference depends on whether we deal with random sampling or randomization. Generalizing inference is about external validity and the uncertainty associated with the sample-based estimation of a population quantity. In contrast, causal inference deals with internal validity and the uncertainty surrounding the randomization-based estimation of a causal treatment effect *within* a given sample of subjects ("sample average treatment effect").
8. While combining both types of inferences and estimating a "population average treatment effect" is often desired, the generalization of causal effects toward a population can only be supported by statistics if, in addition to being randomly assigned to treatments, the experimental subjects had been randomly selected from that population in the first place.

2.5   Recommended Reading

Jeffreys, H. (2011). *Scientific inference* (3rd ed.). Cambridge University Press.
Kalaian, S. A., & Kasim, R. M. (2008). External validity. In P. J. Lavrakas (Ed.), *Encyclopedia of survey research methods* (pp. 254–257). Sage.
Lavrakas, P. J. (2008). Internal validity. In P. J. Lavrakas (Ed.), *Encyclopedia of survey research methods* (pp. 345–351). Sage.
Lohr, S. L. (2019). *Sampling: Design and analysis* (2nd ed.). CRC Press.
Roe, B. E., & Just, D. R. (2009). Internal and external validity in economics research: Tradeoffs between experiments, field experiments, natural experiments, and field data. *American Journal of Agricultural Economics, 91*(5), 1266–1271.

Chapter 3
The Basics of Statistical Inference: Simple Random Sampling

3.1 The Starting Point: Descriptive Statistics of a Given Dataset

For any given set of data—be it a random sample or not—we can compute **summary statistics** that describe certain properties of the data. Describing a large set of observations (e.g., the individual incomes of all sample members) through a summary statistic can provide an informative overview. While a virtually unlimited number of statistics can be conceived of, widely used summary statistics include:

- the (arithmetic) mean of a variable X or Y, usually denoted by $\bar{x}$ and $\bar{y}$
- a variable's standard deviation s and its variance s^2
- the covariance (or, joint variability) $s_{x;y}$ of two variables X and Y, and their correlation $r_{x;y} = s_{x;y}/(s_x \cdot s_y)$
- a difference d between the means of two groups in the sample (also briefly called a "mean difference")
- a regression slope coefficient b (e.g., a linear X-Y-relationship) in the sample, or
- a statistic such as $s/\sqrt{n}$ or its square s^2/n

Statistics such as the mean or the standard deviation that provide a summary description of the observations $x_i, y_i, \ldots$ ($i \in \{1, 2, \ldots, n\}$) of various variables are intuitively meaningful. In contrast, statistics that algebraically relate a variable's dispersion to the size of the sample seem to have less informational value regarding the properties of the given dataset. However, they can become "**inferential statistics**" that have meaning in estimation *if* the sample was randomly drawn from a broader population ("parent population"). In the case of simple random sampling

Supplementary Information The online version contains supplementary material available at [https://doi.org/10.1007/978-3-030-99091-6_3].

(SRS), for example, the statistic $s/\sqrt{n}$ becomes the estimated "standard error of the mean"[1] (see Sect. 3.3.1 for details). This is but another label for the standard deviation of the (sampling) distribution of all sample means that we would find if we *independently* drew very many equal-sized random samples from the *same* population. There are more descriptive sample statistics that can gain inferential meaning. In the case of SRS, for example, important inferential statistics are:

- The sample mean $\bar{x}$ becomes the estimate $\widehat{\mu}$ for the unknown population mean μ, which coincides with the first moment of the sampling distribution (i.e., the average $\bar{\bar{x}}$ of the means found across *very frequently* repeated random draws).
- The corrected sample standard deviation s becomes the estimate $\widehat{\sigma}$ for the unknown population standard deviation σ (see Footnote 1).
- The corrected sample covariance $s_{x;y}$ of two variables X and Y becomes the estimate $\widehat{\sigma}_{x;y}$ for the unknown population covariance $\sigma_{x;y}$, and the sample correlation $r_{x;y}$ becomes the estimate $\widehat{\rho}_{x;y}$ for the unknown population correlation $\rho_{x;y}$.
- The mean difference d between two groups in the sample becomes the estimate $\widehat{\delta}$ for the unknown difference δ in the population, which coincides with the first moment of the sampling distribution (i.e., the average $\bar{d}$ of the differences found across *very frequently* repeated random draws).
- The sample regression coefficient b becomes the estimate $\widehat{\beta}$ for the unknown population regression coefficient β, which coincides with the first moment of the sampling distribution (i.e., the average $\bar{b}$ of the coefficients found across *very frequently* repeated random draws).
- And, as already mentioned, the sample statistic $s/\sqrt{n}$ becomes the estimate $\widehat{SE}$ for the standard error of the mean (see Sect. 3.3.1 for the standard errors of other quantities).

It must be emphasized that all statistics are computed from a particular set of data. They are, therefore, first of all, summary statistics that (hopefully) inform us about interesting properties of the given data. But while statistics never provide an automatic way to assess the validity of findings beyond the confines of the particular sample, they can gain inferential meaning *if* we have a **probabilistic data generation process** such as random sampling. The statistic $s/\sqrt{n}$, for example, can be algebraically computed from any set of values, but it is *not* an estimate of the standard error of the mean unless our set of values is an SRS (see Chap. 4 for information concerning more complex sampling designs).

---

[1] The *corrected* sample standard deviation $s = \left(\frac{1}{n-1} \cdot \sum_{i=1}^{n} (x_i - \bar{x})^2 \right)^{0.5}$, which is the (unbiased) estimator for the population standard deviation, needs to be used.

3.2 Random Sampling, Sampling Error, and Sampling Distribution

In empirical research, we are regularly interested in making inferences in terms of **estimating** population quantities from samples. For example, we might have measured the average per capita income in a sample of 10,000 randomly selected residents, but be ultimately interested in the average per capita income in a city of 1 million residents. To understand the potential and the limits of statistical inference, we must be critically aware of the specific **sampling design** (e.g., SRS, stratified sampling, cluster sampling, etc.) that determines how members of the population are selected into the sample. We also must clearly distinguish between **sample quantities**, which can serve as estimates for population quantities *if* we have a random sample, and the unknown **population quantities** (technically speaking, "population parameters") of interest. Otherwise one runs the risk of lurching all-too readily from the description of some conveniently available data to overconfident generalizations beyond the confines of the particular sample (Lohr, 2019).

Failing to be explicit regarding the data collection design and the **inferential target population** fuels the risk of rash statements regarding real-world states and regularities. Berk and Freedman (2003) emphasize that **statistical assumptions are empirical commitments** and that acting as if one obtained data through random sampling does *not* create a random sample. Ensuring the empirical procedure "random sampling" permits using the standard error as an aid for making generalizing inferences from the sample toward its parent population. Non-compliance, in contrast, implies that all sample statistics, including those that typically carry inferential meaning in random samples (e.g., $s/\sqrt{n}$), remain pure data descriptions. They are devoid of inferential meaning—except when deviations from random sampling are demonstrably ignorable or adequately corrected for (cf. Sect. 4.4).

In brief, statistical inference presupposes **random sampling** and then concerns itself with **random sampling error**, i.e., the fact that even a random sample does not exactly reflect the properties of the population. Because of random sampling error, any sample effect size (e.g., a sample mean $\bar{x}$ or a sample regression slope b) deviates more or less from the true population effect size (e.g., the population mean μ or population regression slope β). Sampling error occurs even when a study is flawless. Due to the **law of large numbers,** it decreases when we make the sample larger; and it would completely disappear if we could study the full population. This is not the case for non-sampling errors such as measurement or model specification errors. They are often hard to detect and might cause deviations from the true value even when we are so lucky as to be able to study the full population.

In many contexts, we will find considerable **sample-to-sample variation** when we repeatedly draw random samples from a population. Therefore, sample-based estimates of population quantities are inherently uncertain (noisy). Even the most rigorous studies concerned with the estimation of a particular population quantity may, therefore, produce highly variable findings. In other words, because we usually cannot study full populations but only estimate population quantities from (random) samples, the scientific enterprise of accumulating evidence from one study to the next is hampered by the "noise" caused by random error. Fortunately, **unbiased estimators estimate correctly on average** over many random samples from the same population. Therefore, one must consider the estimates of all properly implemented studies to obtain an appropriate "meta-proposition" of the population effect size (cf. Chap. 5 for information on meta-analysis). In contrast, an exclusive focus on single studies would hinder knowledge accumulation.

The **uncertainty of an estimation** resulting from random sampling error is quantified by the standard error. The **standard error** describes the standard deviation of the **sampling distribution**, i.e., the distribution of the sample-based estimates that we would obtain across a very large number of equal-sized random samples drawn from the same population.[2] To be more explicit: we study one sample but envisage what would happen if we very frequently drew equal-sized random samples from the same population (statistical replications). That is, we ask how much a sample effect size such as a regression slope b, which we consider as estimate $\widehat{\beta}$ for the unknown population effect size β, would vary around its expected value $\overline{b} = \beta$ across *frequent* statistical replications (cf. the probability density function f in the upper part of Fig. 3.1). Therefore, we speak of "*frequentist* statistics" when we use inferential statistics that are conceptually based on the sampling distribution. A tantamount description of the variation across statistical replications is provided by the lower part of Fig. 3.1. The standard error is now depicted as the standard deviation of the sampling error $e_{\text{sam}} = \widehat{\beta} - \beta$ around its expected value of zero.

It must be noted that the standard error $\widehat{\text{SE}}$ is itself but an estimate of the standard deviation SE of the sampling distribution. As an estimate, it underlies sample-to-sample-variability. To obtain an unbiased estimation of the standard error, we must not only consider the **specific statistic** (mean, mean difference, regression coefficient, etc.) but also the **specific sampling design** (SRS, stratified sampling, cluster sampling, etc.). When we use the correct mathematical formula (see Sect. 3.3 and Chap. 4), the $\widehat{\text{SE}}$ indicates the standard deviation of the sampling distribution to the best of the knowledge that we can have based on the given random sample. Furthermore, we can assume the sampling distribution of the effect size estimates to be approximately normal according to the **central limit theorem**[3] when we have "reasonably large" sample sizes, which is commonly interpreted as $n \geq 30$.

---

[2] No specific label is used for the *variance* of the sampling distribution (i.e., the squared standard error). We must therefore derive from the context whether "variance" describes a variable's dispersion in the sample or the dispersion of the sampling distribution.

[3] While there are many versions, simply put, the central limit theorem states that when we sum up a sufficiently large number of independent random variables, their appropriately scaled sum will approximate the normal distribution, irrespective of the distributions of the single variables (cf. e.g., Rohatgi & Saleh, 2015). The most popular application in statistics refers to the mean and states that, if sample size is sufficiently large (i.e., $n \geq 30$), the sampling distribution of the mean approaches the normal distribution under regular circumstances, irrespective of the variable's distribution in the population. While further assumptions are needed for quantities such as regression slopes (e.g., the least squares assumptions for OLS regressions, cf. Wooldridge, 2020: Chap. 5), we generally rely on the assumption of a normal sampling distribution throughout this book.

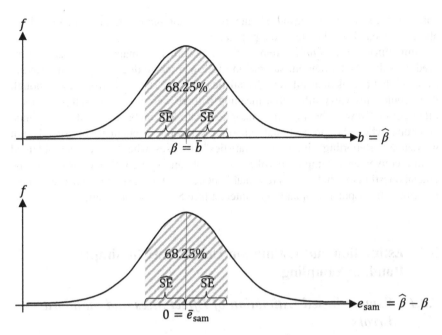

Fig. 3.1 Sampling distribution of a regression slope estimate $\widehat{\beta}$ and distribution of the sampling error $e_{sam} = \widehat{\beta} - \beta$

Unfortunately, in the social sciences, we rarely have random samples. We often use pre-existing observational data or survey data obtained from **convenience samples** that are non-random but easy to come by. Examples are students in classrooms, volunteers on Web-platforms or people recruited in certain venues. This causes serious problems when we want to infer something about a population based only on such a sample. For example, you do not need to be an expert in statistical inference to suspect that an estimate of national voting intention based on a survey of 1000 people emerging from a political party conference may be biased and not as trustworthy as a survey of 1000 people polled at random across the country.

Even when we can avoid **under- and overcoverage** of certain segments of the population and succeed in assembling random subject pools ("sampling frames") from defined populations, we usually do not escape sample **selection bias** because we cannot force subjects to participate in a survey. In longitudinal studies, we additionally face the specific selection problem that study participants may "get lost" over time ("attrition"). When we do not know the mechanism through which members of the population are eventually selected into the sample, we do not know either to what extent certain segments of the population are under- or overrepresented in the sample and to what extent sample members are systematically different from other members of the population. As a consequence, effect size estimates as well as the corresponding standard error estimates will be biased in unknown ways. For example, if, unknown to us, basketball players are overrepresented in a convenience sample, the mean body height in the sample will tell us little to nothing about the mean body height in the population.

Non-random selection is often obvious. Nonetheless, many studies tacitly proceed as if they had a random sample and follow a misguided routine of displaying statistics that are taken as inferential statistics such as "standard errors" even though they would only carry inferential meaning if there were a random sample. This is often associated with the implicit assumption that SRS can be used as default, which is wrong when the proportionate random selection mechanism of SRS is not guaranteed. Reporting "inferential statistics" in cases where data were obtained from a convenience sample provokes inaccurate and overconfident answers to the crucial question of what we can reasonably believe to have learned from a set of data regarding the population quantity of interest (see Sect. 4.4 for details).

3.3   Estimation and Estimation Uncertainty in Simple Random Sampling

3.3.1   Sample-Based Estimation of Effect Sizes and Standard Errors

Simple random sampling represents the most basic case of a random process of data generation that, as such, is the prerequisite for an unbiased estimation of both population effect sizes and standard errors. As we know, the estimate of the standard error quantifies the uncertainty of the effect size estimate ("point estimate") caused by random sampling error. When the data collection design deviates from basic SRS, obtaining unbiased estimates is regularly more complex and sometimes impossible (cf. Chap. 4).

An SRS is a subset of a population drawn with the same probability as any other conceivable subset of identical size, such that each unit had equal probability of being selected into the subset. Since there is no systematic under- or oversampling of certain segments of the population ("**proportionate sampling**"), an SRS represents an unbiased representation of the population. It therefore permits using **unweighted sample statistics** such as a mean $\bar{x}$, a mean difference d or a regression slope b as unbiased estimates of the respective population quantities. SRS also facilitates a relatively easy estimation of standard errors.

Hereafter, we describe how, in an SRS, point and standard error estimates are computed for three of the most common quantities: mean, mean difference (difference between the mean of two groups), and regression slope.

Point Estimate of the Mean The sample mean $\bar{x}$ can be used as unbiased estimate $\hat{\mu}$ for the unknown population mean μ:

$$\bar{x} = \frac{\sum x_i}{n} = \hat{\mu}, \tag{3.1}$$

where x_i are the values of the random variable X observed in the sample, and n the sample size.

The **standard error of the mean** $\bar{x}$ is given by:

$$\widehat{SE}_{\bar{x}} = \left(\left(1 - \frac{n}{N}\right) \cdot \frac{s^2}{n} \right)^{0.5}, \tag{3.2}$$

where n is the sample size, N the population size, and s the corrected sample standard deviation (cf. Sect. 3.1, Footnote 1).

Contrary to standard textbook representations that are based on the implicit assumption that a sample contains only a very small fraction of the parent population, Eq. (3.2) includes the finite population correction factor $1 - n/N$. The **finite population correction** (*fpc*) accounts for the fact that random sampling error not only decreases with growing sample size but also when sample size becomes large *relative* to the population (see Sect. 4.5). For easy traceability, we omit the *fpc* from the following formulas. However, one should recall that having sampled a large portion of the population reduces the standard error. Researchers are advised to use the *fpc* when sample share is more than 5% (Knaub, 2008). When 5% of the population are included in the sample, the *fpc* reduces the standard error by 2.5%. For a share of 50% (75%), the reduction of the standard error increases to 29.3% (50.0%).

Point Estimate of the Mean Difference The sample mean difference d between two groups can be used as unbiased point estimate $\hat{\delta}$ for the unknown mean difference δ in the population:

$$d = \bar{x}_1 - \bar{x}_2 = \hat{\delta}, \tag{3.3}$$

where $\bar{x}_1$ is the mean of observations for variable X in group 1, and $\bar{x}_2$ the mean in group 2.

The **standard error of the mean difference** d between two groups is given by:

$$\widehat{SE}_d = \left(\frac{s_1^2}{n_1} + \frac{s_2^2}{n_2} \right)^{0.5}, \tag{3.4}$$

where s_1^2 and s_2^2 are the corrected standard deviations of the two groups, and n_1 and n_2 their respective sizes.

Point Estimate of the Regression Slope Coefficient Obtaining a *sample* regression slope coefficient that adequately reflects the properties of the given dataset in terms of the association between a variable X (explanatory variable) and a variable Y (response variable) is more complex than obtaining the *sample* mean or the *sample* mean difference. Presupposing SRS *and* a data structure that justifies specifying the regression model as a simple linear regression with a single explanatory variable and an intercept, the sample slope coefficient b can be used as unbiased point estimate $\widehat{\beta}$ for the unknown population regression coefficient β:

$$b = r_{x;y} \cdot \frac{s_y}{s_x} = \frac{s_{x;y}}{s_x^2} = \widehat{\beta}, \tag{3.5}$$

where $r_{x;y}$ is the correlation between variables X and Y, s_x and s_y their respective corrected standard deviations, and $s_{x;y}$ their corrected covariance.

The **standard error of the regression slope coefficient** b in a simple linear regression is given by:

$$\widehat{SE}_b = \left(\frac{1}{n-2}\sum e_i^2\right)^{0.5} \cdot \frac{1}{\sqrt{n}\, s_x}, \tag{3.6}$$

where e_i are the residuals (i.e., the deviations of the dependent variable observations from the estimated regression function), and s_x is the corrected standard deviation of the independent variable X.

To avoid confusion, it must be observed at all times that random error is the only kind of error that the standard error and derived inferential statistics can deal with. Recall that the standard error is but another label for the standard deviation of the (sampling) distribution of the estimates that we would obtain over frequently repeated random draws of equal-sized samples from the same population. Non-random errors such as those caused by non-random sampling or faulty measurements cannot be assessed through inferential statistics.

Conventional regression terminology obfuscates that statistical inference is limited to dealing with the uncertainty surrounding the estimation of a population effect size caused by random error. Regression is about finding the mathematical manipulation of the data that adequately describes relationships between variables in the sample. For example, when the X-Y-relationship in the data is exponential rather than linear, modeling a linear function would be inadequate. Contrary to simpler quantities such as the mean $\bar{x}$, the determination of a X-Y-relationship is often itself called "estimation." This might cause confusion because an identified relationship such as a sample slope b is primarily a description of the relationship in the dataset. Whether the *sample X-Y-relationship* can be used as an estimate for the unknown *population X-Y-relationship* is a critical issue. In a convenience sample, for example, certain segments of the population may be systematically under- or overrepresented. Therefore, sample quantities might be systematically different from population quantities (cf. Sect. 4.4). It may nonetheless be informative to describe interesting properties of the *given* dataset through summary statistics such as a sample mean or a sample regression slope. However, these sample quantities remain sample statistics and do not provide adequate estimates for quantities beyond the non-random sample. Similarly, a dispersion statistic such $s/\sqrt{n}$ can be

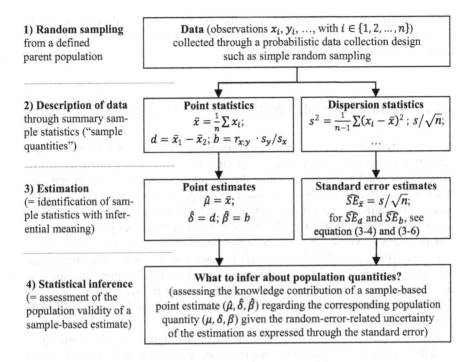

1) Random sampling from a defined parent population

Data (observations $x_i, y_i, \ldots$, with $i \in \{1, 2, \ldots, n\}$) collected through a probabilistic data collection design such as simple random sampling

2) Description of data through summary sample statistics ("sample quantities")

Point statistics
$$\bar{x} = \frac{1}{n}\sum x_i;$$
$$d = \bar{x}_1 - \bar{x}_2; b = r_{x;y} \cdot s_y/s_x$$

Dispersion statistics
$$s^2 = \frac{1}{n-1}\sum(x_i - \bar{x})^2 ; s/\sqrt{n};$$

3) Estimation (= identification of sample statistics with inferential meaning)

Point estimates
$$\hat{\mu} = \bar{x};$$
$$\hat{\delta} = d; \hat{\beta} = b$$

Standard error estimates
$$\widehat{SE}_{\bar{x}} = s/\sqrt{n};$$
for $\widehat{SE}_d$ and $\widehat{SE}_b$, see equation (3-4) and (3-6)

4) Statistical inference (= assessment of the population validity of a sample-based estimate)

What to infer about population quantities? (assessing the knowledge contribution of a sample-based point estimate $(\hat{\mu}, \hat{\delta}, \hat{\beta})$ regarding the corresponding population quantity (μ, δ, β) given the random-error-related uncertainty of the estimation as expressed through the standard error)

Fig. 3.2 The general approach of statistical inference illustrated for SRS

algebraically computed from any set of values, but it is *not* an estimate of the standard error of the mean unless our set of values is an SRS.

Figure 3.2 summarizes the general approach of statistical inference. First, there must be a **probabilistic process of data generation** such as SRS. As a result, we have a sample of size n that contains observations $x_i, y_i, \ldots (i \in \{1, 2, \ldots, n\})$ for the variables of interest. Second, the obtained data can be described through various **summary sample statistics** that inform us about certain properties of the given data. Examples are point statistics such as the mean $\bar{x}$ of all observations for a variable X, the difference d between the means of two groups, and the regression coefficient b that describes the linear relationship between two variables X and Y in the sample. The variance s^2 (or its square root, the standard deviation s), is the usual summary statistic to describe a variable's dispersion around the mean. Third, some of the sample statistics gain inferential meaning and can be used for **estimation** when the data were probabilistically generated. The arithmetic mean $\bar{x}$, for example, becomes a valid estimate $\hat{\mu}$ for the unknown population mean μ if the set of values is an SRS. Similarly, the statistic $s/\sqrt{n}$ becomes a valid estimate of the standard error of the mean $\widehat{SE}_{\bar{x}}$. Fourth, realizing that statistics never provide an automatic way to assess the validity of findings beyond the confines of the particular sample, we engage in **statistical inference**, i.e., we use statistics in a reasoning concerned with the validity (informational value) of the finding and assess what we can reasonably claim to learn from a sample-derived estimate regarding the corresponding population quantity. In this validity assessment, we take into account the uncertainty of the estimation caused by random sampling error.

3.3.2   An Illustrative Application: Gender Pay Gap

We use a stylized example to illustrate what it means to use an observed sample quantity as an estimate for a population quantity of interest. Our presumed research interest is to gain knowledge regarding the magnitude of the pay gap between men and women in Phantasia. We assume that two independent research teams tackle the issue but use different sample sizes (cf. Table 3.1). The first team draws an SRS of size $n = 600$ from all residents of Phantasia. We assume that exactly 300 women and 300 men happen to be sampled. Because the second research team has less financial resources, it draws an SRS of size $n = 60$. For convenience sake, we assume, again, that the random draw results in equal-sized groups of men and women. Imagine now that, by coincidence, both studies find exactly the same mean pay difference $d = 100$ and identical within-group standard deviations $s = 744.5$ in the male and female groups.

Now, both research teams must "make sense of their figures" and draw inferences from the finding in their particular sample toward the population. While both teams found an identical effect size of 100, the variability of the gender pay differences that would be found across repeatedly drawn random samples is much smaller in the large study 1 compared to the small study 2. We estimate this variability according to Eq. (3.4) and obtain $\widehat{SE}_d = 60.788$ for study 1 as opposed to $\widehat{SE}_d = 192.229$ for study 2. With some loss of information, the evidence found in the data in terms of the point estimate (signal) and its uncertainty (noise) can be expressed as a **signal-to-noise ratio** z, which is 1.645 in study 1, and 0.520 in study 2. Despite the identical effect size, the researchers from study 2 would be left with more uncertainty than

Table 3.1   Inferential statistics for a gender pay gap example (stylized numbers)

	Study 1		Study 2	
	Male group	Female group	Male group	Female group
Sample size n	300	300	30	30
Standard deviation s	744.5	744.5	744.5	744.5
Group mean $\bar{x}$	2600	2500	2600	2500
Difference in group means (signal): $d = \hat{\delta}$	100		100	
Standard error of mean difference (noise): $\widehat{SE}_d$	60.788		192.229	
Signal-to-noise ratio: $z = \hat{\delta}/\widehat{SE}_d$	1.645		0.520	
p-value (one-sided)	0.050		0.303	

the researchers from study 1. Therefore, they would have to be more cautious in their reasoning what they can claim to have learned from their small sample in isolation regarding the gender pay gap in Phantasia. This is because, due to the **law of large numbers**, the **inferential leap** is wider from a small sample to the parent population than from a large sample.

While we discuss derivatives of the standard error only in Chap. 6, we already report the so-called *p*-value in Table 3.1, according to widespread practices. Anticipating the detailed description in Chap. 6, we provide a short explanation at this point: when computing the *p*-value, we combine the signal-to-noise ratio obtained from the sample with the assumption of no effect in the parent population ("point null hypothesis"). The one-sided *p*-value then answers the question: what is the *conditional* probability of finding the observed signal-to-noise ratio or even a larger one in random replications *if* we assumed the point null hypothesis (here, no mean pay difference between women and men) to be true in the parent population.

3.3.3 Sample-to-Sample Variability of Point and Standard Error Estimates

It is crucial to realize that statistical inference is based on probability theory and not applicable to convenience samples where the mechanism by which population members are selected into the sample is not known. In this case, sample statistics such as $s/\sqrt{n}$, which can be algebraically computed from any set of data, must not be interpreted as inferential statistics (e.g., as standard error). But even in the ideal case where we have random sampling, effect size estimates ("point estimates") as well as standard error estimates may greatly vary between repeatedly drawn random samples. Due to the law of large numbers, both the accuracy and the reliability (precision) of effect size estimates will increase with the growing sample size. To put it the other way round, small samples increase the **sample-to-sample variability** of both effect size estimates and their corresponding standard error estimates.

Ignoring sample-to-sample variability invites rash conclusions. To provide an intuitive illustration of sample-to-sample variability, we mimic frequently repeated random sampling from populations by carrying out two simulations. In each simulation, we generate 10,000 random samples from a presumed population with a linear relationship $y = 1 + \beta x + e$, with $\beta = 0.2$. The two simulations differ only in their normally distributed error terms $e \sim N(\mu; \sigma^2)$, with $\mu = 0$ and $\sigma = 3$ as opposed to $\mu = 0$ and $\sigma = 5$.[4] The sample size is $n = 50$, with x varying from 0.5 to 25 in equal steps of 0.5. For both the $\sigma = 3$ and the $\sigma = 5$ case, we run OLS regressions for each of the 10,000 samples and record the slope estimates $\widehat{\beta} = b$ and their

---

[4]Rather than simulating a finite population in terms of a very large set of observations from which we could then have repeatedly drawn samples with replacement, we directly simulated the 10,000 random samples of size $n = 50$ because this is less costly on time and storage space.

Table 3.2 Cumulative distribution F for selected z-quantiles based on 10,000 simulated samples ($n = 50$ each) for a presumed relationship $y = 1 + 0.2x + e$

	$e \sim N(0; \sigma^2)$, with $\sigma = 3$			$e \sim N(0; \sigma^2)$, with $\sigma = 5$		
$z = \widehat{\beta}/\widehat{SE}_b$	$F(z)$ [%]	$\widehat{\beta}^{\sigma=3}$	$SE_b^{\sigma=3}$	$F(z)$ [%]	$\widehat{\beta}^{\sigma=5}$	$SE_b^{\sigma=5}$
1.0	0.88	0.065	0.065	15.19	0.107	0.107
1.5	3.18	0.097	0.065	30.11	0.139	0.093
1.645	4.71	0.112	0.068	35.14	0.170	0.104
1.960	8.91	0.131	0.067	47.84	0.228	0.116
2.0	9.64	0.109	0.054	49.49	0.259	0.129
2.5	22.16	0.151	0.060	69.45	0.256	0.102
3.0	41.15	0.167	0.056	85.56	0.341	0.114
4.0	80.83	0.243	0.061	98.35	0.355	0.089

standard errors $\widehat{SE}_b$ according to Eqs. (3.5) and (3.6). Since we use an estimator that perfectly matches the true association in the population, we know that the heterogeneity of the results across the 10,000 samples is the exclusive effect of random sampling error. In other words, from each of the 10,000 random samples, a researcher would estimate a different regression slope $\widehat{\beta}$ and a different standard error $\widehat{SE}_b$ due to the vagaries of random sampling.

Table 3.2 shows the sampling distributions of the slope estimates based on the 10,000 simulated samples—for the $\sigma = 3$ and the $\sigma = 5$ case, respectively. The table displays the cumulative distribution F for eight quantiles of the signal-to-noise ratio z. The slope estimates $\widehat{\beta}$ and standard error estimates $\widehat{SE}_b$ that bring about these ratios are also shown. We see that random sampling error can cause large sample-to-sample variation of slope estimates and, to a lesser degree, standard error estimates. This, in turn, causes considerable variation of the signal-to-noise ratio z. For example, in the $\sigma = 3$ case, only 0.88% of the 10,000 samples yielded a ratio $z \leq 1$, but 19.17% produced a ratio $z > 4$. In contrast, in the $\sigma = 5$ case, 15.19% of samples produced $z \leq 1$, and only 1.65% of samples yielded $z > 4$.

Figure 3.3 visualizes the cumulative distribution of the signal-to-noise ratio z for the $\sigma = 3$ and the $\sigma = 5$ case. The fact that the sample-to-sample variability can be of considerable magnitude causes problems in practical research. Contrary to the 10,000 simulated samples that we used for the numerical determination of the cumulative distribution of z, we would usually have but one sample in practical research. Therefore, we would ignore the extent of the sample-to-sample variability. As a consequence, even after having derived the signal-to-noise ratio for a given sample, we would not know whether the particular sample effect size represents an over- or underestimation of the true effect size in the population. We would not know either how often we under- or overestimate the true effect size in frequently repeated random draws from the population.

To illustrate how little a researcher facing a certain signal-to-noise ratio in a single sample would know, let us take an exemplary look at the dotted lines in Fig. 3.3. In addition, let us look back at selected values displayed in Table 3.2. In the $\sigma = 3$ case,

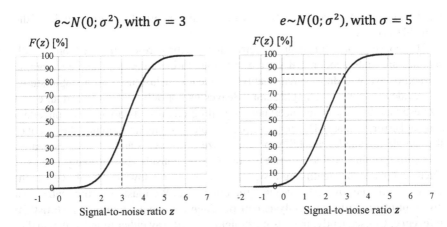

Fig. 3.3 Distribution of the signal-to-noise ratio z over 10,000 samples ($n = 50$ each) for a presumed relationship $y = 1 + 0.2x + e$ in the population

a signal-to-noise ratio $z \leq 3$ was obtained in 4115 of the 10,000 simulated samples; and all of them resulted in an underestimation of the true population effect size $\beta = 0.2$ because even the signal-to-noise ratio $z = 3$ was still only associated with a slope estimate $\widehat{\beta}^{\sigma=3} = 0.167$. In contrast, in the $\sigma = 5$ case, a signal-to-noise ratio $z \leq 3$ occurred in 8556 of the 10,000 simulated samples, but approximately half of them resulted already in an overestimation of the true effect size $\beta = 0.2$. While only selected quantiles are displayed in Table 3.2, this can be seen by looking at the signal-to-noise ratio $z = 1.96$, which was already associated with $\widehat{\beta}^{\sigma=5} = 0.228$. The ratio $z = 3$ was even associated with $\widehat{\beta}^{\sigma=5} = 0.341$. In brief, as practical researchers we would neither know from which population our particular sample was drawn nor from where in the sampling distribution of all conceivable estimates our particular estimate came from. Therefore, inferences tend to be overconfident when too much trust is put on the single study.

While high signal-to-noise ratios are often rashly associated with a high plausibility of estimated effects, the sample-to-sample variability shown in Table 3.2 and Fig. 3.3 reminds us that **unbiased estimators estimate correctly on average** (here, the OLS-estimator). In both populations, we would find an *average* slope across *all* 10,000 simulated samples that is *very* close to the true 0.2 (see Chap. 5 for details on how to use meta-analysis to summarize single studies). But we must realize that we overestimate the effect size *exactly* in those samples that happen to result in very high signal-to-noise ratios. For example, $z = 4$ was accompanied by regression slope estimates $\widehat{\beta}^{\sigma=3} = 0.243$ and $\widehat{\beta}^{\sigma=5} = 0.355$. That is, it is the "abnormally" large sample effect sizes that produce very high signal-to-noise ratios (Trafimow et al., 2018). Consequently, we would inevitably produce an upward bias if we considered only samples with high signal-to-noise ratios, for example, those above a certain threshold such as $z = 1.645$. This is because we would only consider estimates from a left-truncated sampling distribution (cf. Fig. 3.1) instead of using all estimates, as is required for an unbiased assessment of the population effect size in the light of the available evidence.

The consequences of sampling error for practical research based on (small) samples are quite sobering: if we are oblivious to the unknown sample-to-sample variability, we will draw unreasonably confident conclusions from a single study. The evidence found in a single sample leaves us not only with an uncertain effect size estimate but also with uncertainty regarding this uncertainty (i.e., an uncertain standard error estimate). Due to the imponderables of random sampling, the extent of the sample-to-sample variability of effect size estimates, standard error estimates, and corresponding signal-to-noise ratios may be considerable in many contexts. There is no way of identifying a signal-to-noise ratio below which (above which) we underestimate (overestimate) the true effect size. We must therefore realize that **no amount of statistical analysis can produce a formal calculus for statistical inference**. Consequently, there is no straightforward way either to judge the quality of research through a single follow-up study. We need not be surprised if a follow-up study does not even find a similar order of magnitude as the original study, even when both studies are perfectly rigorous.

3.4   Chapter Summary: Statistical Assumptions Are Empirical Commitments

Understanding the potential and the limits of statistical inference in observational studies is straightforward when we remember that it is conceptually based on the sampling distribution and its dispersion. This implies keeping with the following step-by-step approach:

1. We unambiguously define the parent population from which we draw a random sample. The parent population must coincide with the population of interest ("inferential target population") toward which we want to make inferences.
2. We collect a random sample from the defined parent population following a clearly defined probabilistic sampling design (e.g., SRS, stratified sampling, cluster sampling) that exactly defines how members of the population are selected into the sample.
3. We observe an effect size in the sample, for example, a mean, a mean difference or an association between two variables such as a regression slope. Aiming to make generalizing inferences (inductive inferences), we use the observed sample quantity as a point estimate for the population quantity of interest.
4. When assessing the validity of the estimate, we consider that inductive inferences are inherently uncertain and that they do not automatically flow from sample statistics—whatever they may be.
5. While the validity of findings beyond the confines of a single study is impaired by more than just random sampling error (e.g., measurement errors, model specification errors, non-random sampling error), we realize that statistical inference is limited to considering estimation uncertainty resulting from random error.

6. Accounting for the specific quantity of interest (e.g., mean, mean difference, regression slope) and the specific sampling design (e.g., SRS, stratified sampling, cluster sampling), we correctly estimate the standard error to quantify the uncertainty of the point estimation caused by random error.

7. We now possess information that can be understood as "signal" and "noise"—with the sample-based point estimate representing the signal, and the standard error estimate representing the noise from random sampling.

8. We use these two intelligible pieces of information—the point estimate (signal) and its uncertainty (noise)—in a comprehensive scientific reasoning that makes reasonable inductive inferences from the idiosyncratic random sample toward the population of interest in the light of all available information.

To avoid confusion, a final terminological comment must be made: saying that estimation *uncertainty* is used to assess the *validity* of generalizations would be at odds with measurement terminology. Measurement theory distinguishes *precision/ reliability* from *accuracy/validity*. In large samples, for example, we might face little uncertainty and find nearly identical sample effect sizes across statistical replications (precise/reliable estimation). At the same time, they might all be heavily biased estimates of the population effect size due to a faulty measurement instrument, such as a misunderstood survey question (inaccurate/invalid measurement). The terminological conflict would be that, in statistical inference, we somewhat loosely say that the standard error is used to assess the validity of generalizations toward the target population (population validity) while it is but a measure of the uncertainty caused by random (sampling) error.

3.5 Recommended Reading

Berk, R. A., & Freedman, D. A. (2003). Statistical assumptions as empirical commitments. In T. G. Blomberg & S. Cohen (Eds.), *Law, punishment, and social control: Essays in honor of sheldon messinger* (2nd ed., pp. 235–254). de Gruyter.

Hirschauer, N., Grüner, S., Mußhoff, O., & Becker, C. (2018). Pitfalls of significance testing and *p*-value variability: An econometrics perspective. *Statistics Surveys, 12*, 136–172.

Hirschauer, N., Grüner, S., Mußhoff, O., Becker, C., & Jantsch, A. (2021b). Inference using non-random samples? Stop right there! *Significance, October*, 20–24.

Knaub, J. (2008). Finite population correction (fcp) factor. In P. Lavrakas (Ed.), *Encyclopedia of survey research methods* (pp. 284–286). Sage.

Levy, P. S., & Lemeshow, S. (2008). *Sampling of populations. Methods and applications* (4th ed.). Wiley.

Lohr, S. L. (2019). *Sampling: Design and analysis* (2nd ed.). CRC Press.

Rohatgi, V. K., & Saleh, A. K. M. E. (2015). *An introduction to probability and statistics* (3rd ed.). Wiley.

Chapter 4
Estimation Uncertainty in Complex Sampling Designs

4.1 Overview of Different Sampling Designs

There is a vast variety of procedures to draw samples from parent populations. These procedures, which are known as sampling designs, survey designs or data collection designs, differ in the mechanisms they use to select members of the parent population into the sample. The particular properties of the respective selection mechanism must be considered when making generalizing inferences from sample quantities toward population quantities. Figure 4.1 provides an overview of the most important designs. The most basic distinction is between random and non-random sampling. In random sampling, we know how members of the population are probabilistically selected into the sample. In contrast, non-random sampling denotes a convenient or haphazard sample generation where we ignore the probabilities with which population members are selected into the sample. In addition, the special case of full population data is to be considered.

We know from Chaps. 2 and 3 that generalizing statistical inference presupposes random sampling and then deals with random sampling error. Without random sampling, there is no random sampling error and, therefore, neither need nor room for assessing random-error-related uncertainty. However, random sampling is but an umbrella term for a vast variety of very different probabilistic designs. Their respective features must be considered when choosing the mathematical formulas for the estimation of population quantities (e.g., mean, mean difference, regression slope) and their standard errors.

So far, we have focused on simple random sampling (SRS) as the most basic probabilistic selection mechanism (cf. Chap. 3), which serves as a benchmark because each member of the parent population has equal probability of being selected into the sample. Besides SRS, it is useful to distinguish two more basic sampling designs: stratified sampling (cf. Sect. 4.2) and cluster sampling (cf. Sect. 4.3). In addition, there is non-random or convenience sampling (cf. Sect. 4.4) where no probabilistic information exists how the sample was composed from the parent

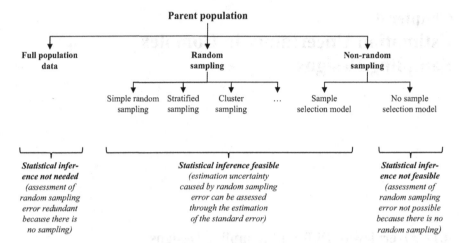

Fig. 4.1 Overview of different sampling designs

population. This leads to contamination with selection bias and rules out the use of inferential statistical procedures (cf. Sect. 4.4.1). The arguable exception is when we have enough information about the non-sample members of the population to run a so-called sample selection model that would enable us to correct for sample selection bias, i.e., the fact that members and non-members of the sample might be systematically different (cf. Sect. 4.4.2). Finally, there is the case of full population "sampling" (cf. Sect. 4.5). When we have full population data, there is no random sampling error and inferential statistical procedures are meaningless unless researchers resort to the often questionable notion of a "superpopulation."

We describe the standard error formulas that must be used in different sampling designs and show that an unbiased estimation of effect sizes and standard errors presupposes a random process of data generation. We focus on the mean because the notation for standard errors of other quantities (e.g., regression slopes) tends to become messy. The fundamental fact, however, that statistical inference presupposes a random process of data generation and that deviations from SRS need to be considered applies also to those quantities. The crucial message is: interpreting summary statistics of a given dataset (e.g., $s/\sqrt{n}$) as inferential statistics is only possible when these probabilistic conditions are met. Statistical inference is precluded when the basic precondition "independent random sampling" is violated.

4.2   Stratified Sampling

Researchers often deviate from SRS because other sampling designs are statistically or financially more efficient. In stratified sampling, we first divide a population of size N into H mutually heterogeneous but internally as homogeneous as

possible subpopulations (or, "strata"). For example, we might divide a population into several income classes (strata). We then randomly sample from *each* stratum h ($h \in \{1, 2, \ldots, H\}$). The simplest case is **proportionate stratified sampling** where we sample an identical fraction of each stratum ($n_1/N_1 = n_2/N_2 = \ldots = n_H/N_H$), ensuring that stratum sample size n_h is proportional to stratum size N_h. In dispro-portionate stratified sampling, in contrast, we oversample certain strata intention-ally—for example, those that exhibit more variability than others in order to reduce the estimation uncertainty caused by random sampling error.

Contrary to proportionate stratified sampling, where each member of the popu-lation has equal probability of being included into the sample, we need to consider weights when a sample is systematically unrepresentative of the population, such as in **disproportionate stratified sampling**. In other words, we need to "reconfigure the sample as if it was a simple random draw of the total population" (Friedman, 2013). The **weight** $w_{hi} = w_h = N_h/n_h$ that is assigned to a sampled unit i ($i \in \{1, 2, \ldots, n_h\}$) in stratum h is the reciprocal of the probability that this unit is included in the sample. It indicates how many units of the full population are represented by a sampled unit. If we sample a 10% fraction in stratum 1 and a 20% fraction in stratum 2, then each sampled unit in stratum 1 has weight $w_1 = 10$ (represents 10 units), whereas each sampled unit in stratum 2 has weight $w_2 = 5$ (represents 5 units). When estimating a population regression from such a sample, we need to apply, in the simplest case, weighted least squares instead of ordinary least squares to the sample regression to obtain unbiased point estimates.

Stratified sampling needs also to be considered when estimating sampling variability. Compared to an SRS of size n, stratified sampling, where $n = \sum n_h$, reduces sampling error. The **estimated standard error of the mean**, for example, is given by (cf. Lohr, 2019: 79):

$$\widehat{\text{SE}}_{\text{Strat}} = \left(\sum \frac{N_h^2}{N^2} \cdot \frac{N_h - n_h}{N_h} \cdot \frac{s_h^2}{n_h} \right)^{0.5}, \tag{4.1}$$

where s_h^2 is the sample variance *within* each stratum h.

When estimating the standard error for stratified samples, we consider the within-strata variance but *not* the variance between strata because we *independently* draw a random sample from *each* stratum. This is why, compared to SRS, the reduction of the standard error is the more pronounced the smaller the variance within the strata and the greater the variance between them. Equation (4.1) uses a finite population correction $(N_h - n_h)/N_h = 1 - n_h/N_h$. This correction considers that sampling error not only decreases with growing stratum sample size but also when stratum sample size becomes large *relative* to stratum size. Since the correction applies *within* each stratum, it can be used to make stratified sampling still more efficient: the **standard error is reduced when strata with high variability are oversampled**. Such oversampling can also be seen as an intentional use of low weights for strata with high variability. Using the definition $w_h = N_h/n_h$ from above, this can be shown by rewriting Eq. (4.1) as:

$$\widehat{SE}_{Strat} = \frac{1}{N} \cdot \left(\sum N_h \cdot (N_h - n_h) \cdot \frac{s_h^2}{n_h} \right)^{0.5} = \frac{1}{N} \cdot \left(\sum w_h \cdot (N_h - n_h) \cdot s_h^2 \right)^{0.5} \quad (4.2)$$

In brief, stratified sampling is a special case of random sampling. Not only are the probabilistic pre-conditions for estimating standard errors met, but standard errors can be adjusted downwards. If we neglected this downward adjustment and used the standard error for an SRS as default, we would base our inferential reasoning on inflated standard errors; i.e., we would be too conservative in our judgments.

4.3   Cluster Sampling

Cluster sampling is popular because randomly sampling from preselected clusters is usually cheaper and more convenient than randomly sampling from the full population. Cluster sampling bears superficial similarity to stratified sampling because it also subdivides a population of size N into exhaustive subpopulations (segments). However, it uses a hierarchical approach to data collection. Earmarking the fundamental difference in the sampling design, we use G to denote the number of segments that are now called "clusters." The difference to stratified sampling is that, instead of randomly drawing units from each segment of the population, we draw, in a primary step, a random sample of g clusters from the population of G clusters, which are called "**primary sampling units**" (PSUs). In a secondary step, we randomly select units, which are called "**secondary sampling units**" (SSUs), from the previously selected clusters. Both stages of data collection need to be accounted for when estimating population quantities and standard errors. Lohr (2019: 168) warns that "[o]ne of the biggest mistakes made by researchers using survey data is to analyze a cluster sample as if it were an SRS. Such confusion usually results in the researchers reporting standard errors that are much smaller than they should be."

One-stage cluster sampling is a simple variant of cluster sampling where *all* observational units (SSUs) within the selected clusters (PSUs) are surveyed. When this is too costly, **two-stage cluster sampling** is applied where we draw an SRS from each selected cluster. A frequently used form of cluster sampling is two-stage area sampling. Imagine we are interested in estimating per capita income in a country with 50 provinces. In area sampling, we might first randomly select ten provinces. In the second step, we would then draw random samples of residents from only these ten provinces. Analogous to stratified sampling, we need to consider **weights** when estimating population quantities from cluster samples that are systematically unrepresentative of the population. The appropriate weight is still the reciprocal of a unit's probability of being included into the sample. However, this probability now derives from the probability that a cluster is selected in the first stage of sampling, *and* the (conditional) probability that a unit within a selected cluster is sampled in the second stage.[1]

Whereas stratification decreases sampling error compared to SRS, the opposite is generally true for cluster sampling. Lohr (2019: 166) illustrates why: "Members of the same cluster tend to be more similar than elements selected at random from the whole population—[...]; residents of the same nursing home tend to have similar opinions of the quality of care. These similarities usually arise because of some underlying factors that may or may not be measurable—[...]. Thus, we do not obtain as much information about all nursing home residents in the United States by sampling two residents in the same home as by sampling two residents in different homes, [...]." Cluster sampling increases the sampling error even when we use equal selection probabilities for all SSUs that facilitate a "self-weighting" point estimation analogous to an SRS.[2]

There are two approaches of obtaining equal probabilities for all observational units (SSUs): (1) We can use probabilities proportional to cluster size when selecting clusters (PSUs) and then sample the same *absolute number* of SSUs in each PSU. Alternatively, (2) we can use a fixed probability for selecting PSUs and then sample an identical *fraction* of SSUs in each PSU (one-stage clustering is a special case of this approach). Imagine a population of size 2000 is divided into 5 large PSUs à 200 and 10 small PSUs à 100. Approach (1) implies that sampling a large PSU is twice as likely as sampling a small PSU (e.g., 0.2 vs. 0.1). An equal number of SSUs (e.g., 50) is then sampled in each PSU—corresponding to a 25% fraction in large PSUs and a 50% fraction in small PSUs. The resulting probability of SSUs being selected would be 0.05, both for SSUs in large PSUs ($0.05 = 0.2 \cdot 0.25$) and for SSUs in small PSUs ($0.05 = 0.1 \cdot 0.5$). In approach (2), a selection probability of 0.05 for all observational units (SSUs) could be obtained, for example, by using an equal probability of 0.1 for selecting PSUs and sampling an identical 50% fraction of SSUs within each selected PSU. While both approaches are self-weighting, approach (2) is expected to produce a larger sampling error (Lohr, 2019: 220).

In **one-stage cluster sampling** with clusters of equal size N/G that are selected with equal probability, estimating the **standard error of the mean** is straightforward (cf. Lohr, 2019: 171):

---

[1] For dealing with cluster samples in the regression context, adequate methods account for between-cluster and within-cluster variability. Keywords in this context are hierarchical, multilevel, random effects or mixed effects models; see McCulloch et al. (2008) and Rabe-Hesketh and Skrondal (2021) for methodological details.

[2] Cluster sampling usually increases standard errors because units in natural clusters such as geographical areas often share environmental influences that make them more homogeneous than units that are randomly selected from the population. We would therefore find positive intra-cluster correlations. The opposite (i.e., cluster members being less homogeneous than randomly selected units) is unlikely to occur in natural clusters (Lohr, 2019: 173ff).

$$\widehat{SE}_{\text{Clust\_1}} = 1/\frac{N}{G} \cdot \left(\left(1 - \frac{g}{G}\right) \cdot \frac{s_t^2}{g}\right)^{0.5},$$ (4.3)

where s_t^2 is the *between*-PSU variance of cluster totals.

Equation (4.3) shows that in one-stage clustering, the standard error depends on the between-PSU variance but not the within-PSU variance. This is because there is no within-PSU sampling error when we observe *all* SSUs in the preselected PSUs.

Even when we maintain the assumption of equal-sized clusters that are selected with equal probability, the formula for estimating the **standard error of the mean** becomes more complex in **two-stage cluster sampling**: we now need to consider the between-PSU *and* the within-PSU variance because sampling error results from two sources: from sampling g clusters in the first stage *and* from sampling n/g units within each selected cluster in the second stage (cf. Lohr, 2019: 185):

$$\widehat{SE}_{\text{Clust\_2}} = 1/\frac{N}{G} \cdot \left(\left(1 - \frac{g}{G}\right) \cdot \frac{s_t^2}{g} + \frac{1}{g^2} \cdot \sum_{i \in \text{sampled clusters}} \frac{N^2}{G^2} \cdot \left(1 - \frac{n/g}{N/G}\right) \cdot \frac{s_i^2}{n/g} \right)^{0.5},$$ (4.4)

where s_t^2 is the *between*-PSU variance of cluster totals, s_i^2 the within-PSU variance of sampled observations from the ith sampled cluster, and n the total sample size.[3]

If we analyzed the data as if they were obtained from SRS, we would mistakenly estimate the standard error according to Eq. (3.2). Basing the estimation of standard errors in a cluster sample—be it for a mean or some other quantity such as a regression coefficient—on the SRS assumption[4] generally leads to an underestimation. This can be quantified by the so-called **design effect**, i.e., the ratio between the correct sampling variance (i.e., the squared standard error) of the cluster sample and the sampling variance of an equal-sized SRS (Kish, 1965: 161). Often, correct standard errors are several times larger than the default standard errors that presume SRS. While the design effect tells us by how much we would underestimate the variance of the sampling distribution if we erroneously used the SRS formula, "it is not a way to avoid calculating variances: You need an estimate of the variance from the complex design to find the design effect" (Lohr, 2019: 309).

---

[3] When *all* clusters are selected in the first stage, we, again, have stratified sampling. Aligning the notation ($g = G = H$, $N/G = N_h$, $n/g = n_h$, $s_i^2 = s_h^2$), Eq. (4.4) can correspondingly be reduced to Eq. (4.1).

[4] It should be noted that even in an SRS design, we must use "heteroscedasticity-robust" standard errors when the dispersion of observations is different in different segments of the population (cf. Wooldridge, 2020: Chap. 8).

Excursus: Besides the survey design literature that shows how to consider cluster sampling, there is a large body of literature on "cluster-robust standard errors" which assumes that, irrespective of the survey design, standard errors need to be adjusted for clusters when the data exhibit cluster-like properties. See Cameron and Trivedi (2005: 829–845), Cameron and Trivedi (2009: 306–311 and 323–329), Cameron and Miller (2015) or MacKinnon (2019). The identification of correlations between individuals within geographical areas such as provinces or villages is a simple example. While the aggregation level (here, villages or provinces) at which one should "cluster" is debated, Cameron and Miller (2015) recommend to search for cluster-like data properties and then to be conservative and use the cluster-robust standard errors resulting from the most aggregate level. The routine of estimating cluster-robust standard errors based on ex post identified cluster-like data structures was recently challenged by Abadie et al. (2017). They claim that the need for clustering adjustments cannot be derived from data properties but must be based on the data generation process. They also note that many datasets do not meet the crucial assumption of cluster-robust standard error estimation that the number of observed clusters should be small relative to the number of clusters in the population. An extreme example to the opposite where all "clusters" are observed illustrates the issue. Imagine a researcher ignores that the data under study resulted from stratified sampling from provinces. Next, observations are found to be similar within the provinces (intra-province correlations). The provinces are then treated as clusters and a cluster-robust estimation obtains standard errors that are larger than conventional standard errors from SRS. But they should be smaller because, in the stratified sampling design, units were sampled from *each* province.

In brief, cluster sampling is another special case of random sampling where the probabilistic pre-conditions for estimating standard errors are met. However, cluster samples usually produce larger sampling errors than SRSs. Hence, standard errors need to be adjusted upwards. If we neglected this upward adjustment and used the standard error for an SRS as default, we would base our inferential reasoning on wrongly deflated standard errors; i.e., we would be too confident and, even worse, be too confident to an *unknown degree*, in our judgments.

4.4 Convenience Samples Contaminated by Selection Bias

4.4.1 Non-randomness: The Big Challenge in the Social Sciences

Observational data are often tacitly analyzed as if they were obtained through random sampling even when a **non-random selection mechanism** was at work.

Such approaches can be seriously flawed. Trafimow (2019: 344) unmistakably warns that "it is the rare study that uses sampling that is completely random and where each participant sampled is independent of the other participants sampled. Without making assertions about exactly how often these assumptions are importantly violated, there can be little doubt that the violation prevalence is considerable. In that case, no inferential statistical procedures, [. . .], are completely valid. [. . .] The notion that inferential statistical procedures may sometimes, and even often, be inappropriate, may be tough medicine for reviewers and editors to swallow. But the medicine nevertheless is therapeutic. Another option is to use methods for addressing violations of independence or random sampling [. . .], while keeping in mind that these have their own assumptions and drawbacks."

Using data from non-probability **convenience samples** is common in empirical studies in the social sciences (Miller, 2017). For example, researchers often ask people who happen to be present in certain venues to participate in a survey. In the social sciences, the most notorious instance are students from Western, Educated, Industrialized, Rich, and Democratic (WEIRD) countries (cf. Henrich et al., 2010) who happen to be in a particular researcher's classroom. Other examples are passers-by in shopping malls, users of social media, and people who happen to be on some mailing list or who explicitly agree to be included in data bases that assemble volunteers for surveys. There are many Web-platforms, for example, where respondents can be recruited at low costs from non-probability panels of individuals who volunteer to participate in surveys for money. Amazon's Mechanical Turk and YouGov are prominent examples.[5]

There are two main reasons why convenience samples might not match parent populations and be systematically unrepresentative: **(1) Over- and undercoverage** (or even non-coverage) means that certain segments of the population are over- or underrepresented in (or even omitted from) the sampling frame in the first place. **(2) Self-selection** describes the fact that people can usually choose whether to respond to a survey or not. This also includes attrition, which is a specific selection problem in longitudinal studies when study participants "get lost" over time.

When we have coverage and selection problems, we do not know the mechanism through which units are selected into the sample. That is, contrary to deliberately disproportionate sampling designs, we ignore to which extent certain population segments are over- or undersampled. Since the selection into the sample may depend on properties that are correlated with the quantity of interest, those who respond may be systematically different from those who do not. We might plausibly expect, for

---

[5]In 2017, Public Opinion Quarterly published a Special Issue "Survey Research, Today and Tomorrow" that tackled the issue of non-probability samples. Its editor notes: "To survive, surveys need to maintain their 'value proposition.' One important part of this proposition is that surveys gather information that is generalizable from the measured sample to a well-defined population. [. . .] There have been prominent clashes between advocates for the probability and nonprobability approaches. Those who support the probability sampling method observe that they have probability theory to support their population inferences, while the advocates for volunteer panels rely on inductive modeling that has no comparable theoretical foundation" (Miller, 2017: 205–207).

example, that (very) rich people are underrepresented among the click workers on Web-platforms. We are thus likely to get a biased view of the mean and the dispersion of incomes when we study a convenience sample of paid volunteers. When the selection of units into the sample is not random, sample quantities may tell us little about population quantities. We must therefore realize that we run the risk of misestimating population quantities and their associated standard errors to a considerable but unknown extent unless we adequately correct for the selection bias that results from the violation of independent random sampling (cf. Rosenthal & Rosnow, 2009: book 3; Vogt et al., 2014; Lohr, 2019).

4.4.2   Approaches to Correct for Selection Bias in Convenience Samples

Approaches to correct for selection bias are best understood by looking back on deliberately disproportionate sampling designs. In these designs, we ex ante knew the selection probabilities and the ways in which the data generating process was systematically different from a simple random draw. We therefore also knew how to "reconfigure the sample" to make it comparable to a simple random draw and ensure that the variable distributions in the sample match the distributions in the parent population. When the composition of the sample is influenced by unrepresentative coverage and/or self-selection, we do *not* know these probabilities ex ante. We only know *that* (certain types of) people might be included into the sample with differing probabilities, depending on individual characteristics (age, gender, education, income, etc.) that may be observable or not. **Sample selection models** are used to ex post control for these selection probabilities. They are based on the assumption that we know all confounding variables which might play a role in subjects' participation in a survey and that we have observations for these variables, both from participants *and* non-participants (**no unmeasured confounding**).

From the late 1970s, an extensive literature developed that dealt with selection problems. Initially, they focused on experimental research where the identification of causal relationships is compromised when randomization is not fully able to balance confounders across groups, for example, due to dropouts after randomization or quasi-experimental designs. Later, their scope of application was expanded to survey research where the unbiased estimation of population quantities is impaired when random sampling is not warranted, for example, due to non-coverage and/or non-response. In the last two decades, a particular focus was put on **selection problems in Web surveys** and, methodologically, on **propensity score models**, which were imported to the survey context from experimental research.[6]

---

[6] For further details, see Heckman (1979), Berk (1983), Heckman et al. (1997), Luellen et al. (2005), Rosenbaum (2010), Austin (2011), Brick (2013), Valliant et al. (2013), Guo and Fraser (2014), Imbens and Rubin (2015), and Mercer et al. (2017).

Propensity score models are probably the most intuitive way to address selection bias. In experimental research based on randomization, propensity score models focus on causal inferences within the group of experimental subjects (internal validity). In this context, they are used to deal with the problem of unbalanced confounders across treatment groups. Mercer et al. (2017: 253) explicitly liken selection processes in survey research and experimental research: "A probability-based survey is essentially a randomized experiment where the pool of subjects is the set of units on the sampling frame and the treatment is selection into the survey. Unlike experiments where we observe outcomes on both treated and untreated subjects, in surveys we observe outcomes only on the selected units, with the expectation that there should be no difference between selected and non-selected units."

In the survey context, propensity score models ideally use all selection-relevant variables to estimate subjects' probability (the "propensity score") of participation as opposed to non-participation (Heckman et al., 1997). Similar to the ex ante known selection probabilities in disproportionate sampling, these ex post estimated probabilities are used to reconfigure the sample by correcting for any imbalances between those who are in the final sample and those who are not. Propensity scores are therefore also called "balancing scores," *conditional* on which the distribution of (potential) confounders in the participation group is assumed to match the distribution in the non-participation group (Austin, 2011).

Let us look at a stylized example to illustrate how propensity score methods work in principal. Imagine we are interested in the mean per capita income in a population of $N = 10,000$ subjects. Assume also that we luckily had access to a sampling frame in the form of a truly random pool of 1000 subjects (500 men, 500 women) and that we addressed a survey to the entire pool. The response rate was only 15%, resulting in a sample size of $n = 150$. In this sample, we find 100 men (20% of the men on the sampling frame), but only 50 women (10% of the women on the sampling frame). We observe a mean income in the sample of $\bar{x} = 6$. Taking a closer look, we find a mean of $\bar{x}_m = 7$ among the 100 men, and a mean of $\bar{x}_f = 4$ among the 50 women. It now seems natural to equate the gender-specific participation shares with selection probabilities ("propensity scores") and act on the assumption that men selected themselves into the sample with a propensity of 0.2 ($= 100/500$), and women did so with a propensity of 0.1 ($= 50/500$). To correct for the fact that gender is not equally distributed across participants and non-participants (oversampling of men, undersampling of women), each man in the sample would have to have a weight of $w_m = 5$ and represent five men, whereas each woman would have to have a weight of $w_f = 10$ and represent ten women. We could now use the weights-corrected sample mean $\bar{x}_w = 5.5 = (5 \cdot 100 \cdot 7 + 10 \cdot 50 \cdot 4)/1000$ as estimate for the mean income in the population.

In this highly stylized example, we assumed that gender was the only variable affecting self-selection into the sample. To correct for selection bias, we could, therefore, use a very simple propensity score model that corresponds to simple **post-stratification**. That is, we could put the presumably ex post detected self-selection probabilities on the same level as intentional selection probabilities that we as researchers might have used ex ante in a deliberately disproportionate sampling design. In our stylized example, the same formulas as in the stratification case apply and Eq. (4.1) can be used to estimate the standard error of the mean.

In reality, many variables other than gender (e.g., age, education or even income itself) might also influence the probability of survey participation. For example, residents with high incomes might be less inclined to respond. As a consequence, simple post-stratification will not suffice to account for the biased sample composition and to adequately estimate population parameters and standard errors (cf. de Leeuw et al. (2008: 317–341) for standard error estimation under various versions of weight correction). Similar to cluster sampling, but contrary to our stylized example above, the consideration of self-selection usually causes standard errors to be larger than those that would be obtained from analyzing the sample as if it were an SRS (Copas & Li, 1997). Still worse, the demanding data requirements of sample selection models can often not be met in practical research. For example, if age, gender, education, and income affect participation, we need to know the distribution of these variables not only among participants but also among non-participants.

Excursus: The need and the possibility to correct for self-selection are related to the notion of "missing data" (cf. Mercer et al., 2017). In SRS, no corrections are needed because data are missing completely at random (MCAR), meaning selection into the sample is unconditionally independent of the variable of interest (no confounding). Somewhat difficult to grasp at first, the label "data missing at random" (MAR) is used when selection bias can be "remedied" by making unit selection independent of the variable of interest conditional on observed confounders (no unmeasured confounding). Using the propensity scores of men and women to correct for the confounder "gender" was a simple example. When not all confounders are known or can be observed, it is impossible to adequately correct for selection bias (unmeasured confounding). This corresponds to the notion of "data missing not at random" (MNAR), which rules out the use of inferential statistics.

While the formal presentation of propensity score models or other sample selection models (cf. e.g., Little & Rubin, 2020) is beyond the scope of this book, it should be recognized that we need to be generally wary of miss-specifying the selection model. When groups with distinct characteristics are completely missing in the sample or when we do not know, or are not able to measure, all selection-relevant variables, we cannot properly correct for selection bias. While emphasizing that correction methods should be used whenever sufficient data are available, Cuddeback et al. (2004: 22) note that "failing to use these methods when appropriate is different than failing to use them when data for non-participants are unavailable, which is common. In this latter case, sample selection models are obviously of no use." Of course, researchers can never be completely sure that they know all selection-relevant variables. And often, the necessary data from non-participants will be lacking. Even when considerable data is available for non-participants, one should remain skeptical whether all confounding variables that affect selection were considered. Going beyond conventional routines, this needs to be explicitly reflected in the interpretation of results to avoid overconfident inferential reasoning.

In brief, we often use convenience samples because we are unable to comply with the "empirical commitment" of random sampling. Even when we follow the *procedure* of random sampling, the *result* will often not be a random sample because of uncontrolled self-selection. Ignoring whether and how a sample was probabilistically composed from a defined population rules out estimating standard errors because no sampling distribution can be envisaged. In some cases, we may have enough information from non-participants to correct for selection bias, which rehabilitates the probabilistic foundations for using inferential statistics (Levay et al., 2016). When we are not able to perform corrections, which is likely to be more often the case in practical research than the literature implies, we should transparently communicate that sample statistics have no inferential meaning.

Inductive inferences may still be feasible even when we do not have a random sample. They must be based on substantive arguments outside statistics, however. For example, we might find a harmful effect of a hitherto unknown drug in an accidental group of people (convenience sample). Even though we might completely ignore the group's composition in relation to the population and know nothing about its representativeness, we might infer that the reactions to the drug are likely to be similar in other people. This conclusion would be based on the physiological argument that human organisms are quite comparable across people. While the scope of such a conclusion may be debated among medical experts, we must not forget that scientific inferences toward populations cannot be supported by statistical arguments when there is no random process of data generation (random sample). When we ignore the (probabilistic) mechanism through which population members are selected into the observed sample, all sample statistics remain purely descriptive summary statistics that carry no statistical inferential meaning.

4.5   Full Populations and Finite Population Correction

Random sampling error occurs even when studies are flawless. But due to the law of large numbers, the uncertainty of an estimation resulting from random sampling error decreases when sample size increases. We might also say that the **inferential leap from the data to the population** decreases when there are more observations. Random sampling error completely disappears when we can study full populations. That is, there is neither need nor room for statistical inference when we already have data for an entire population.

Vogt et al. (2014: 243) note that reporting *p*-values in the analysis of full population data is nonetheless quite common—for example, when analyzing aggregated economic data. They also provide an illustrative example of a study that looks at the association between mean education levels and birth rates in *all* UN member states. Since no sample-to-population inference is necessary, interpreting summary data statistics as inferential statistics does not make sense. This intuitive fact is formally reflected in the **finite population correction** (*fpc*) factor $1 - n/N$, which is used for adjusting the *squared* standard error (cf. Sect. 3.3.1). Instead of implicitly assuming that a sample was drawn from an infinite population—or at least that a small sample of size n was drawn from a very large population of size N—the *fpc* considers that sampling error decreases not only with growing sample size but also when the fraction of the population that is sampled becomes large.

Researchers are advised to use the *fpc* when a sample share is greater than 5% (Knaub, 2008). When 5% of the population is sampled, the *fpc* reduces the standard error by 2.5%. For a share of 50%, the reduction increases to 29.3%. Having data for an entire population ($n = N$) results in an *fpc* of zero, which leads to a corrected standard error of zero. This is consistent because there is no random sampling error when the "sample" covers 100% of the target population. In contrast, *non-random errors* can cause serious validity problems even when we are in the comfortable position of having access to full population data. Non-random errors, such as measurement and model specification errors, often impair inductive inference far more than the idiosyncrasies of random samples. They are not probabilistically accessible and cannot be dealt with by inferential statistical procedures.

The *fpc* is often ignored and "standard errors" are displayed even in the case of **full population data** (100% samples). To justify this, one must somehow introduce the notion of a sampling error and a sampling distribution. This implies *imagining* an infinite "unseen parent population" ("**superpopulation**") and a stochastic data generating process from which one has presumably observed one random realization (Hirschauer et al., 2020b). In so doing, the observed population becomes in imagination a random sample of a (parent) superpopulation. Even back in the 1980s, Denton (1988: 167) noted that this is a rhetorical device (also known as "**great urn of nature**") that does not evoke wild enthusiasm from everybody. However, some random process of data generation—and not just a record of empirical data—has to be presumed for inferential statistical procedures to make any sense.

When we have observations for a full population, adopting the notion of a superpopulation might arguably be useful in some research contexts. For example, researchers could imagine a **quasi-stable real-world system** whose operative regularities, while being possibly limited to a finite time interval, exceed the time interval over which the population data were collected. In the above UN example, this would imply envisaging structural regularities that are reflected in a stochastic data generating process which not only produced the association between education and birth rates in *all* UN member in the period for which the data were collected but that will also be at work in the next period(s). Obviously, the plausibility of such an assumption and thus the meaningfulness of generalizing claims based on inferential statistics can only be assessed when the claim is openly made and the source of uncertainty clearly articulated, and when the time interval from which the data were taken as well as the time interval over which the data generating system is presumably at work are clearly defined.[7]

Sometimes an imaginary population and an imaginary random sampling mechanism are even assumed in the case of **non-random samples** taken from a larger population of interest. Assume, for example, a researcher addresses a survey to the 100 students who happen to be in the classroom on a particular Monday morning. We know that such a convenience sample might be seriously biased and tell us little about a broader population of interest. Assuming in such a context that there is a random process of data generation does not facilitate statistical inference in any conventional sense of the word. Instead, it requires **considering the sample as the relevant finite population**. That is, instead of generalizing toward a numerically larger parent population of students, "inferences" would be limited to an unseen superpopulation in terms of a random generating process that is valid for *exactly* these 100 students. No statistical inferences beyond the sample of the 100 students can be made (Hirschauer et al., 2020b) without running a sample selection model based on sufficient information about the population (cf. Sect. 4.4.2). Of course, this raises the question of whether we can learn something from an unfathomable fiction of an imaginary population and an imaginary random sampling mechanism when we have only a convenience sample.

Berk and Freedman (2003: 238–239) challenge the belief that imaginary populations and imaginary sampling mechanisms can produce the statistical equivalent of a random sample: "In order to use the imaginary-population approach, it would seem necessary for investigators to demonstrate that the data can be treated as a random sample. [. . .] We doubt the case could be made for the [. . .] example or any similar illustration. Nevertheless, reliance on imaginary populations is widespread. Indeed, regression models are commonly used to analyze convenience samples: as we show later, such analyses are often predicated on random sampling from imaginary populations. The rhetoric of imaginary populations is seductive precisely because it seems to free the investigator from the necessity of understanding how data were generated." This is a dangerous fallacy because, as this chapter has unmistakably demonstrated, it is essential to consider the data generation process when estimating effect sizes and standard errors.

In brief, when researchers display inferential statistics even though they have data from full populations or even convenience samples, they face the logical necessity to resort to an imaginary superpopulation and random sampling mechanism. Even if one accepted this twofold fiction, one would have to be clear that it would completely change the inferences compared to conventional random samples taken from well-defined parent populations. Rather than just routinely displaying statistics that delusively suggest conventional inferential content, researchers should clearly

---

[7]Stochastic time series analysis is a field to which the notion of a superpopulation is inherent. Stochastic time series analysis presumes that observations that are made sequentially over time are but realizations (i.e., a sample) of an infinite number of time series (i.e., the superpopulation) that the underlying stochastic data generating process could have produced.

communicate when they resort to this fiction. That is, they should explicitly discuss why and how they base their inferential reasoning on the notion of a superpopulation. This is but a specification of the **general desideratum that researchers clearly describe the sampling process and the population of interest** from which the sample comes and to which they want to generalize.

4.6   Chapter Summary: Inference Requires Considering the Sampling Design

This chapter focused on the often overlooked fact that the very pre-conditions for using inferential statistics are not met when the sampling design is ignored or when studies are based on non-probability samples (violation of independent random sampling). When there is no random sampling, no random sampling error and no sampling distribution can be envisaged. Random data generation is a *necessary* condition for an unbiased estimation of population parameters and a meaningful use of standard errors. When data do not satisfy this probabilistic requirement, all sample statistics remain purely summary statistics that describe certain properties of the given dataset but have no statistical inferential content.

Critically reflecting on the conceptual pre-conditions for using inferential statistics is important. It calls to mind the nature of statistical inference, which—even if applicable—solely deals with the uncertainty of estimates caused by random error, and which is therefore only a part of scientific inference. Furthermore, it (hopefully) generates awareness that the assumption of independent random sampling is violated in convenience samples because they exhibit one of the two following drawbacks (or both): first, the units on the sampling frame are conveniently chosen but not probabilistically selected from the target population. Due to such coverage problems, there are unknown differences between the units who are on the sampling frame and those who are not. Second, the units that happen to be on the sampling frame select themselves with unknown probability into the final convenience sample which, if not corrected for, leads to self-selection bias. While these two features often occur jointly in practical research, each of them suffices to rule out a meaningful use of inferential statistics.

To ensure successful scientific communication, non-probability samples must be clearly distinguished from probability samples: (1) When researchers succeed in obtaining probability samples, they should transparently state which sampling design was used, from which population the sample was drawn, and consequently, to which population they want to make inductive inferences. Without providing such basic information, inference statements are nebulous, at best—and misleading, at worst. (2) When researchers are limited to using non-random samples that preclude the use of inferential statistics, they should be clear about it and refrain from displaying statistics that are generally taken to carry inferential meaning. Displaying "inferential" statistics in circumstances where population inferences cannot be supported by probability theory is likely to cause overconfident conclusions. Alas, it still seems a widespread spontaneous reflex among researchers who often do not explicitly question whether there is a chance model upon which to base statistical inference.

4.7 Recommended Reading

Berk, R. A., & Freedman, D. A. (2003). Statistical assumptions as empirical commitments. In T. G. Blomberg & S. Cohen (Eds.), *Law, punishment, and social control: Essays in honor of sheldon messinger* (2nd ed., pp. 235–254). de Gruyter.

Hirschauer, N., Grüner, S., Mußhoff, O., Becker, C., & Jantsch, A. (2021b). Inference using non-random samples? Stop right there! *Significance, October*, 20–24.

Hirschauer, N., Grüner, S., Mußhoff, O., Becker, C., & Jantsch, A. (2020b). Can *p*-values be meaningfully interpreted without random sampling? *Statistics Surveys, 14*, 71–91.

Knaub, J. (2008). Finite population correction (fcp) factor. In P. Lavrakas (Ed.), *Encyclopedia of survey research methods* (pp. 284–286). Sage.

Levy, P. S., & Lemeshow, S. (2008). *Sampling of populations. Methods and applications* (4th ed.). Wiley.

Little, R. J. A., & Rubin, D. B. (2020). *Statistical analysis with missing data* (3rd ed.). Wiley.

Lohr, S. L. (2019). *Sampling: Design and analysis* (2nd ed.). CRC Press.

Mercer, A. W., Kreuter, F., Keeter, S., & Stuart, E. (2017). Theory and practice in nonprobability surveys. Parallels between causal inference and survey inference. Special issue 2017: Survey research, today and tomorrow. *Public Opinion Quarterly, 81*(S1), 250–271.

Miller, P. V. (2017). Is there a future for surveys? Editorial to the special issue 2017: Survey research, today and tomorrow. *Public Opinion Quarterly, 81*(S1), 205–212.

Rosenbaum, P. R. (2010). *Design of observational studies*. Springer.

Vogt, W. P., Vogt, E. R., Gardner, D. C., & Haeffele, L. M. (2014). *Selecting the right analyses for your data: Quantitative, qualitative, and mixed methods*. The Guilford.

Chapter 5
Knowledge Accumulation Through Meta-analysis and Replications

5.1 The Basics of Meta-analysis

5.1.1 Dealing with Different Measurements and Model Specifications

Economic analyses are often based on multiple regressions in which a dependent (or, response) variable (usually denoted by Y) is modeled as a function of several independent (or, explanatory) variables (usually denoted by X_j). The independent variables are often divided into focal variables of interest (*focal predictors*) and variables that are used to control for confounding influences (*control variables*). The essential outcome of regression analysis are the sample regression slopes b_j, which relate the independent variables X_j to the dependent variable Y. When the sample is a random sample, the sample regression slopes b_j can be used as estimates $\widehat{\beta}_j$ for the unknown population slopes β_j. If several studies have tackled the same X-Y-relationship (e.g., between education X and income Y), we are naturally interested in summarizing these findings to make more trustworthy inferences toward the parent population than we could from a single study.

In the scientific exercise of inductive reasoning (generalizing inference), we generally face the task of drawing conclusions from the particular (e.g., a **sample quantity**) to the general (a **population quantity**), which is usually a real-world state or relationship of interest. The education-income relationship is one example. There are many others: we might use sample data to estimate how large the pay gap is between men and women in a certain country. Other questions of interest might be the relationship between the exposure to the herbicide Glyphosate and the incidence of cancer, or the link between the use of pesticides and earthworm populations in

Supplementary Information The online version contains supplementary material available at [https://doi.org/10.1007/978-3-030-99091-6_5].

N. Hirschauer et al., *Fundamentals of Statistical Inference*, SpringerBriefs in Applied Statistics and Econometrics, https://doi.org/10.1007/978-3-030-99091-6_5

agricultural soils. In the attempt of arriving at reasonable generalizations, we must be aware of the fundamental limitation of inductive reasoning. This limitation is that we cannot obtain certain knowledge regarding a broader context from a particular set of data. Even when these data were obtained through random sampling, **each random sample is idiosyncratic**; i.e., it has its own properties that do not exactly reflect the properties of the parent population. Due to sample-to-sample variability, sample-based estimates of population quantities are inherently uncertain. Therefore, we must embrace any remaining uncertainty, assess its magnitude, and remain cautious in our judgments of what we should most reasonably believe in the light of the available evidence.

Due to the **law of large numbers**, the uncertainty of estimates caused by random sampling error decreases the larger the size of the sample or the larger the number of samples that we analyze. We might also simply say that the **inferential leap** from the given data toward the population becomes smaller when we have more observations. This is obvious when we look back at Eq. (3.6) which gives the estimate for the standard error of the slope coefficient in a simple linear regression. An exaggerated focus on a single study based on limited sample size causes cognitive biases in terms of overconfident conclusions that do not adequately consider sample-to-sample variability. Borenstein et al. (2009: xxi) state that "rather than looking at any study in isolation, we need to look at the body of evidence." This raises the question of *how* we should summarize the findings of individual studies to obtain an appropriate picture of the state of knowledge that is available in a given field of research. Under certain conditions, **meta-analysis** can be used to arithmetically synthesize the statistics of individual studies into summary **meta statistics** (cf. Sect. 5.1.2). Analogous to single studies, the crucial meta statistics are the point estimate, which is now the "summary effect size," and its estimated uncertainty, which is now the "summary standard error."

The most basic precondition for a meaningful synthesis of different studies is their ability to produce knowledge about the **same unknown population quantity**. Imagine several studies are concerned with an education-income relationship. Assume that these studies use similar regression models but have drawn random samples from different populations (e.g., France, Ghana, Russia, and the US). Since each single study is based on data from, and concerned with, a different *national* education-income relationship, computationally summarizing the regression coefficients from the different studies does not make sense. While its meaningfulness would be arguable, a different meta-analytical perspective would arise if we considered these national populations as a (random) cluster sample (cf. Sect. 4.3) of all nations in the world. In this case, the unknown population quantity of interest would be the *global* education-income relationship. And similar to cluster sampling (cf. Sect. 4.3), meta-analysis would have to consider the between-study variance (analogous to the between-PSU variance) as well as the within-study variance (analogous to the within-PSU variance) to obtain an adequate summary standard error (cf. Borenstein et al., 2009: Chap. 12).

Even when different studies are concerned with the same quantity and population, computationally summarizing their effect sizes is not meaningful when the individual estimates are not comparable. Programmed research, where a superordinate meta-planner ensures that a series of studies are consistently carried out using the same measurements and research design, is the only situation where meta-analysis (i.e., the arithmetic summary of weighted effect size estimates) is without problems. However, some technical deviations from this ideal can be accounted for. **Different metrics** (scales), for example, rule out a direct comparison of the absolute ("raw") estimates from different studies. But converting among scales may be easy in some cases. A simple example is the conversion from one **response scale**, such as 1–5, to another scale, such as 1–10 or 1–100. More sophisticated approaches must be used to make **mathematically different metrics** comparable. An example would be a study that reports a log odds ratio whereas another one a mean difference (for more details, see Borenstein et al., 2009: Chap. 7).

In some instances, the scales that are used in different studies to measure a certain theoretical construct cannot be converted into each other because they result from **substantively different measurements**. One of many examples is when different elicitation procedures are used to measure people's risk attitudes. In such cases, **standardized effect sizes** such as Cohen's d are suggested (see Borenstein et al. 2009: 25ff). Instead of using the raw mean difference d according to Eq. (3.3), **Cohen's** d measures the difference $\bar{x}_1 - \bar{x}_2$ between the mean of two groups in units of the groups' joint dispersion s_{pool} ("pooled standard deviation"):

$$\text{Cohen's } d = \frac{\bar{x}_1 - \bar{x}_2}{s_{pool}}, \text{ with } s_{pool} = \left(\frac{(n_1 - 1)s_1{}^2 + (n_2 - 1)s_2{}^2}{n_1 + n_1 - 2}\right)^{0.5} \quad (5.1)$$

Comparison problems are particularly pressing issues when we want to synthesize regressions slopes from different studies because the slope estimates depend on the scales that are used for the predictor variable X and the response variable Y. Standardized effect size measures are also available for regression slopes. The fact that slope estimates are usually denoted by $\widehat{\beta}$ might cause confusion because standardized regression slopes are often called "**beta-coefficients**." In simple linear regression, the beta-coefficient (here denoted by $\widehat{\beta}_{beta}$) is obtained by multiplying the regression slope $\widehat{\beta}$ with the ratio of the standard deviation of X and Y. Using Eq. (3.5), we see that $\widehat{\beta}_{beta}$ coincides with the correlation $r_{x;y}$ between X and Y:[1]

$$\widehat{\beta}_{beta} = \widehat{\beta} \cdot \frac{s_x}{s_y} = r_{x;y} \cdot \frac{s_y}{s_x} \cdot \frac{s_x}{s_y} = r_{x;y} \quad (5.2)$$

A large variety of standardized effect size measures have been suggested for various research contexts. But while they might be the best we can do under

---

[1] The computation of the slope coefficient $\widehat{\beta}$ according to Eq. (3.5) applies only in the case of a simple linear regression. Therefore, $\widehat{\beta}_{beta} = r_{x;y}$ applies also only to simple linear regression. However, the general form $\widehat{\beta}_{beta} = \widehat{\beta} \cdot s_x/s_y$ applies also to multiple linear regression.

circumstances that otherwise rule-out meta-analytical approaches, all standardized effect sizes share a major drawback: we lose any possibility of assessing the substantial size and economic importance of an effect. Take the example of the gender pay gap. Equation (5.1) shows that Cohen's d will be small even in the case of large pay gaps when the within-group standard deviations are also large. Similar problems arise when we use the "**common language effect size**," which has been suggested as an alternative to Cohen's d (McGraw & Wong, 1992). The common language effect size is a version of the Wilcoxon-Mann-Whitney statistic. It indicates the probability that a randomly selected member of the first group (here, men) shows a larger outcome value (here, pay) than a randomly selected member of the second group (here, women). Again, the information regarding the effect size is lost, but we would be able to compare studies even if they used very different measures of pay such as monthly or hourly wages or even a relative measure of pay compared to a certain benchmark.

Another problem for meta-analysis arises from **different model specifications**. This may be especially pronounced when studies concerned with the same question but originating from different disciplines (e.g., economics and psychology) are to be summarized. For illustration sake, let us look at a simple example. Imagine one education-income study uses a measure of "social background" as a control variable, whereas another one does not. Ideally, this should not occur because all studies should use adequate model specifications. This includes the requirement that the explanatory variables in the model match those in the parent population. Nonetheless, identical model specifications are rarely found across non-programmed individual studies, even when they tackle the same research question. Similar to different measurements, heterogeneous model specifications cause an apples-and-oranges problem for meta-analysis because, as we know from experience, the inclusion of controls into a regression may substantially influence the effect size estimate. In some contexts, **meta-regression** may provide a solution to the apple-and-oranges problem (cf. Borenstein et al., 2009: Chap. 20). Simply speaking, meta-regression treats the estimated effect sizes across the summarized studies as dependent variable that is modeled as a function of the true effect size and a variety of individual study characteristics that are used as control variables.

In brief, the comparability of effect size estimates across studies is restricted by some critical features. This holds especially for the social sciences. One reason that often makes meta-analysis difficult if not impossible is the fact that social sciences research is predominantly a bottom-up exercise. As such, it produces a large quantity of empirical results on topical issues but is plagued by an enormous **heterogeneity of measurement approaches and model specifications** (Bruns, 2017). This heterogeneity creates a large obstacle to systematic research syntheses. Consequently, quantitative meta-analyses are rare, and it is difficult to get a clear view of the state of the knowledge even when they are made. This is due to technical restrictions as well as researchers' flexibility in deciding which studies to include into the synthesis. Even a cursory look at economic publications shows that a related criticism that Stanley and Jarrell (1989: 162) directed to literature reviews is still not overcome: "The reviewer often impressionistically chooses which studies to include in his review, what weights to attach to the results of these studies, how to interpret the results, [...]. As a result, reviews are rarely persuasive to those who do not already number among the converted."

5.1.2 Synthesizing Effect Sizes and Standard Errors Across Several Studies

In what follows, we use a stylized case that resembles a programmed research context where a whole series of commissioned studies use the same research design. We use this simple case to illustrate the basic concept of meta-analysis. A thorough grasp of the meta-analytical concept will deepen the understanding that each properly implemented study makes a valuable knowledge contribution. This holds even though sample-based estimates may vary considerably from study to study because of random sampling error. Understanding the logic of the meta-analytical approach will also help readers internalize the universal desideratum that we consider the body of evidence instead of asking the question of whether a particular study found the "true" effect or not. The requirement to consider the whole body of evidence holds even when formal meta-analysis is not feasible.

We use the **univariate weighted least squares** approach to demonstrate the fundamentals of meta-analysis. We assume that we are to summarize 20 studies based on different sample sizes ($n = 20, 30, 40, 50$). For easy intuition, we avoid all complications, such as different metrics between studies. Mimicking frequently repeated random draws, we use simulation to generate the 20 samples from a presumed population where $y = \alpha + \beta x + e$, with $\alpha = 1$, $\beta = 0.2$, $e \sim N(0; \sigma^2)$, with $\sigma = 5$, and $x \in \{25/n, 2 \cdot 25/n, 3 \cdot 25/n, \ldots, (n-1) \cdot 25/n, 25\}$. In each of the 20 studies, an OLS regression is used to estimate the intercept $\widehat{\alpha}$ and the slope coefficient $\widehat{\beta}$.

Using the weighted least squares method, the summary coefficient $\widehat{\beta}_{\text{sum}}$ that synthesizes the coefficients of the single studies is computed as (cf. Becker & Wu, 2007: 7):

$$\widehat{\beta}_{\text{sum}} = \sum_{m=1}^{M} \widehat{\beta}_m \cdot w_m / \sum_{m=1}^{M} w_m, \text{ with } w_m = 1/\widehat{\text{SE}}_m^2, \qquad (5.3)$$

where M is the number of studies in the meta-analysis, and $\widehat{\beta}_m$ the slope estimate in the mth study.

The weight w_m that is attributed to the slope estimate of each study m is the reciprocal of its squared standard error estimate $\widehat{\text{SE}}_m^2$; and the ratio $w_m / \sum_{m=1}^{M} w_m$ denotes the percentage weight of each study. The standard error estimate $\widehat{\text{SE}}_{\text{sum}}$ of the summary coefficient $\widehat{\beta}_{\text{sum}}$ is computed as:

Table 5.1 Meta-analysis for 20 single studies based on simulated random samples from a presumed population where $y = 1 + 0.2x + e$; $e \sim N(0; \sigma^2)$, with $\sigma = 5$

Study no. m	Observations n per study	Estimated coefficient $\widehat{\beta}$	Standard error $\widehat{SE}$	Signal-to-noise ratio $z = \widehat{\beta}/\widehat{SE}$	Weight (%) $w/\sum w$
1	20	−0.078	0.161	−0.487	2.53
2		−0.012	0.146	−0.083	3.07
3		0.348	0.166	2.094	2.37
4		0.110	0.153	0.718	2.80
5		0.434	0.155	2.799	2.72
6	30	0.280	0.118	2.370	4.68
7		0.203	0.131	1.550	3.83
8		0.204	0.131	1.564	3.84
9		0.225	0.149	1.506	2.94
10		0.410	0.121	3.401	4.50
11	40	0.171	0.113	1.509	5.10
12		0.209	0.127	1.644	4.05
13		0.264	0.092	2.853	7.66
14		0.248	0.108	2.300	5.62
15		0.077	0.091	0.843	7.83
16	50	0.180	0.101	1.783	6.41
17		0.219	0.088	2.497	8.53
18		0.284	0.102	2.791	6.33
19		0.251	0.089	2.811	8.21
20		0.221	0.097	2.286	6.99
Meta-analysis of all 20 studies (total no. of observations: 700)		0.216	0.026	8.450	100.00
Single large regression (over all 700 observations)		0.201	0.027	7.553	–

$$\widehat{SE}_{sum} = \left(1/\sum_{m=1}^{M} w_m\right)^{0.5} \tag{5.4}$$

Table 5.1 displays the regression slope coefficients $\widehat{\beta}$ and their standard errors $\widehat{SE}$ as estimated in the 20 single studies that were subjected to the meta-analysis. The second-last column shows the signal-to-noise ratio z for each study. The last column gives the percentage weight of each study. The second-last row at the bottom of the table shows the results of the meta-analysis: the summary coefficient $\widehat{\beta}_{sum}$, its standard error $\widehat{SE}_{sum}$, and the summary signal-to-noise ratio.

Since we know all raw data, we can also carry out a single large regression over all $n = 700$ observations. Looking at the second-last and the last row of the table, we see that meta-analysis worked quite well. The computed summary coefficient $\widehat{\beta}_{sum} = 0.216$ is slightly above the true effect size $\beta = 0.2$. The single large regression performed even better and obtained a slope estimate $\widehat{\beta}_{700} = 0.201$, which is very close to the true effect size. The differences between the standard error estimates $\widehat{SE}_{sum} = 0.026$ and $\widehat{SE}_{700} = 0.027$ are small.

Several general conclusions can be derived from this highly stylized example of meta-analysis based on the univariate weighted least squares approach:

1. Meta-analysis is capable of leaving behind the exclusive focus on the single study and its exaggerated preoccupation with a single signal-to-noise ratio.
2. The signal-to-noise ratio provides neither an indication of whether a study is valuable nor whether a study's estimate is close to the real-world quantity.
3. Meta-analysis includes the contribution to knowledge that each single study makes even when it is so small that its individual estimate is associated with a lot of uncertainty. It thus considers that the effect best supported by a given set of data is always the observed effect, regardless of the signal-to-noise ratio.
4. By considering the body of evidence instead of single studies in isolation, a good estimate can be obtained with little remaining uncertainty. In our example, the summary slope $\widehat{\beta}_{sum} = 0.216$ approximates $\beta = 0.2$ quite well and it is associated with a very small uncertainty ($\widehat{SE}_{sum} = 0.026$), which, in turn, leads to a very large summary signal-to-noise ratio of more than 8.
5. We would inevitably produce an upward bias if we considered only a selection of studies with high signal-to-noise ratios z. For example, if we included only studies with $z \geq 1.645$ into the meta-analysis, we would obtain a substantially upward-biased summary slope estimate of 0.266.

Meta-analysis is based on the crucial fact that proper sample-based estimation requires unbiased estimators and that **unbiased estimators estimate correctly *on average***. Therefore, one must jointly consider the estimates of *all* properly implemented studies to obtain the most appropriate and reasonable proposition regarding the real-world state of interest. In our example, this includes the very small estimates, such as $\widehat{\beta} = -0.078$ (study 1), which even has an opposite sign, as well as the very large estimates, such as $\widehat{\beta} = 0.434$ (study 5). It cannot be overstressed that the necessity to consider all studies applies even though, from our superior knowledge of the real $\beta = 0.2$, we might rashly dismiss these extreme estimates as "wrong." But in the scientific process of accumulating evidence from a sequence of studies, they are as valuable and necessary as the other estimates in that we need them all to obtain an appropriate picture of the real effect.

To prevent excessive expectations, it must be stressed that basic meta-analytical approaches cannot control for distortions resulting from improper single studies ("**garbage-in-garbage-out**") or faulty selection mechanisms. In our stylized example, we did neither have a garbage-in-garbage-out problem nor a faulty selection mechanism: all studies used the same (OLS) estimator that perfectly matched the true association; and no study was excluded from meta-analysis based on a faulty criterion such as a minimum signal-to-noise ratio. Prevalent publication practices often cause serious deviations from this ideal because studies with high signal-to-noise ratios are taken to be more trustworthy. They are, therefore, also more likely to be published.

The distortion of the body of evidence toward results associated with high signal-to-noise ratios is called **"publication bias"** (or "file drawer problem"). Unfortunately, publication bias is rather the rule than the exception (for more details, see Sect. 6.2.4). Basic meta-analytical methods such as weighted least squares are as such not able to control for such distortions because they simply summarize the studies that are included into the analysis. **In the case of bias, meta-analysis will summarize the bias.**[2]

5.2 Evaluation of the Quality of Research Through Replications

The inevitability of sample-to-sample variability should make us wary of asking the delusive question of whether a particular follow-up study is a successful replication of a prior study or not. Recall that the 20 studies described in Table 5.1 were all perfectly rigorous, but that they obtained estimates within the very wide range from $\widehat{\beta} = -0.078$ to $\widehat{\beta} = 0.4344$. In the meta-analysis, we needed both extremes as well as all other 18 estimates in-between to obtain an appropriate picture of the real-world relationship $\beta = 0.2$. In other words, even when looking only at these two studies, we would not convey useful information by speaking of a replication failure. We must avoid such statements because even such widely divergent estimates can be due to random sampling error, in which case we have to take into account their respective knowledge contribution and include them in meta-analysis.

The description of meta-analysis underpins the necessity to consider the whole body of evidence provided by rigorous studies, irrespective of their results. However, the question is left open of how we should scrutinize the quality of prior research when we carry out a replication *exactly because* we fear that the claims of an original study might be dubious. Using methods such as meta-analysis, which, by design, would impartially synthesize the estimates from the original study and its replication, would not make sense: in the case of bias, meta-analysis summarizes the bias. How, then, should we scrutinize the quality of prior research?

The answer regarding an ostensibly desirable criterion for an unambiguous distinction between "replication success" and "replication failure" is sobering: nothing can be said for sure about *one* individual study based on the result of *one* single follow-up study. Based on the results only (i.e., without scrutinizing the

---

[2] Meta-analytical approaches dealing with bias in the published body of evidence are beyond the scope of this book. For an overview of methods to gauge reporting and publication biases, see Hirschauer et al. (2018). For the further study of meta-analysis, the reader be referred to Becker and Wu (2007), Borenstein et al. (2009), Card (2012), Kline (2013: Chap. 9), Schmidt and Hunter (2014), and Andrews and Kasy (2019).

methodological approach), we cannot identify whether the original study was biased, or whether the follow-up study was biased, or whether differing outcomes are simply due to the variability caused by random sampling error. This holds especially when the follow-up study uses a similar sample size. It holds to a lesser degree when the replication study carries (much) more evidential weight because it uses a (much) larger sample. As we know, this is because the uncertainty of estimates caused by random error decreases with growing sample size due to the law of large numbers. Carrying out a *very* large replication—or very many replications with moderate sample size—would yield an estimate of the population quantity with little uncertainty. We could use such a relatively "precise" estimate as a pertinent benchmark with which we could compare the estimate of a scrutinized study.

We often do not have enough financial resources to do a *very* large replication for each of a large number of original studies. When resources are limited, there are two meaningful, cautious approaches to replications and their evaluation:

1. We can content ourselves with scrutinizing the overall quality of a body of research in a certain field or journal—rather than trying to judge individual studies. For example, we could limit ourselves to doing one replication for each of a large number of agricultural economics studies. We could then analyze how well we can replicate the effect size estimates *on average* across all those studies. While we would not be able to judge any single study, we would rightly question the quality of agricultural economics research if follow-up studies in this field found substantially smaller effect size estimates on average compared to those of the original studies.

2. A cautious approach to replication on the level of the individual study is to search for a provisional assessment of whether the results of the original and the replication study corroborate each other. This assessment is facilitated by the fact that, across repeated random draws, more samples yield estimates close to the true population quantity than far from it (cf. Fig. 3.1). In addition, we know that we do not have to worry about differences between estimates that are too small as to be substantially relevant. Therefore, when the replication study comes up with an effect size estimate of similar magnitude as that of the original one, we can speak of the results as being in line (consistent) with each other.

However, as mentioned above, we should better avoid attaching the label "replication success" or "replication failure" to a follow-up study. We must not forget that substantially similar as well as substantially differing estimates could simply be due to the vagaries of random sampling. Especially the label "replication failure" might thus wrongly suggest to many that a follow-up study that comes up with dissimilar effect size estimates is an indication of a low quality of the original research and a false finding. To prevent such premature, spontaneous misinterpretations, we generally avoid these normatively connoted labels. Instead, we stick to simply describing the result of a follow-up study as being substantially similar to the original one or not. This is in line with Sohn's (1998: 292) emphasis on the question of whether a replication study produces "the essentially same outcome as that of the original study." Even more pointedly, Hubbard et al. (2019: 91) note that "[a] major

Table 5.2 Two replication contexts for an original gender pay gap study

	Sample size: men + women	Standard deviation s	Estimated effect size $\hat{d}$	Estimated standard error $\widehat{SE}$	Signal-to-noise ratio $z=\hat{d}/\widehat{SE}$
Replication context 1					
(a) Study 1 (original)	300 + 300	744.5	100	60.79	1.645
(b) Study 2 (follow-up)	30 + 30	744.5	100	192.23	0.520
Replication context 2					
(a) Study 1 (original)	300 + 300	744.5	100	60.79	1.645
(c) Study 2 (follow-up)	30,000 + 30,000	744.5	10	6.08	1.645

focus of scientific inference can be viewed as the pursuit of significant sameness, [...]." The judgment of what represents "**significant sameness**" is not within the reach of statistics. Instead, it needs to be based on expert judgment of what is a "substantial difference" in the particular field of interest. For example, a monthly gender pay gap of 1 euro would probably not qualify as a substantial and economically relevant effect in Germany, whereas a 1-Euro-gap in hourly wages would.

We take up to the gender pay gap example and two stylized replication settings to illustrate what is meant by significant sameness and consistent results. The upper part of Table 5.2 reproduces the gender pay gap example described in Table 3.1. Study 1, which is based on a sample of 600 people from the population in Phantasia, is now understood to be the original study. Study 2, which uses a sample of 60 people from Phantasia, is understood to be a follow-up study that was only able to draw a small sample because very limited resources could be mustered for the replication. For convenience sake, we assume, again, equal-sized groups of men and women and identical within-group standard deviations. Since both studies are based on SRS, Eqs. (3.3) and (3.4) are used to estimate the mean pay differences and associated standard errors.

In the original study, the estimate of the mean pay difference (signal) of $\hat{d} = 100$ exceeds the standard error (noise) of $\widehat{SE} = 60.79$ by a factor $z = 1.645$. We assume that by sheer coincidence, the follow-up study finds the same mean pay difference $\hat{d} = 100$. However, due to the small sample, the uncertainty of the estimation as quantified through the standard error is much larger in the small follow-up study compared to the original study. Given the identical effect size estimate, this results in a considerably smaller signal-to-noise ratio $z = 0.520$. If we were reduced to seeing only the evidence from the small study 2, we would have to be very cautious in our inferences regarding the mean gender pay gap in Phantasia. It is obvious, however, that even though the signal-to-noise ratio in the second study differed considerably from the ratio in the first study, we found significant sameness (consistent results) in both studies: after all, both studies found an *identical* difference of 100. The take-

Table 5.3 Constellations of significant sameness and non-sameness for a mean difference

	Substantially similar effect ("significant sameness")	Substantially dissimilar effect ("significant non-sameness")
Similar uncertainty (standard error)	(1) $\widehat{d}_1 \cong \widehat{d}_2$ and $\widehat{SE}_1 \cong \widehat{SE}_2 \rightarrow z_1 \cong z_2$	(3) $\widehat{SE}_1 \cong \widehat{SE}_2$ and ... $\widehat{d}_1 \gg \widehat{d}_2 \rightarrow z_1 \gg z_2$ $\widehat{d}_1 \ll \widehat{d}_2 \rightarrow z_1 \ll z_2$
Dissimilar uncertainty (standard error)	(2) $\widehat{d}_1 \cong \widehat{d}_2$ and ... $\widehat{SE}_1 \gg \widehat{SE}_2 \rightarrow z_1 \ll z_2$ $\widehat{SE}_1 \ll \widehat{SE}_2 \rightarrow z_1 \gg z_2$	(4) $\widehat{SE}_1 \gg \widehat{SE}_2$ or $\widehat{SE}_1 \ll \widehat{SE}_2$ and $\widehat{d}_1 \gg \widehat{d}_2$ or $\widehat{d}_1 \ll \widehat{d}_2$ $\rightarrow z_1 \cong z_2$ or $z_1 \not\cong z_2$

home message that should not be forgotten is: being a quotient, the value of z depends on the effect size estimate *and* its uncertainty (standard error) which, in turn, crucially depends on the sample size. Due to its mathematical construction as a quotient of the effect size and its standard error, the signal-to-noise ratio z is *not* able to provide any indication of whether a replication produced significant sameness or not (cf. Greenland et al., 2016).

Things are different in replication context 2. Here, the original study is followed by a replication study that is based on a *very* large sample of size $n = 60,000$. The mean pay difference estimated in this follow-up study is only $\widehat{d} = 10$. Due to the very large sample size and the small remaining uncertainty as quantified by the standard error $\widehat{SE} = 6.08$, we can reasonably claim that the magnitude of the pay gap in the population is in the one-digit or very low two-digit area. That is, the estimated mean pay difference $\widehat{d} = 10$ in the obviously more reliable study 2 is substantially different from the estimate $\widehat{d} = 100$ of the original study. Significant sameness is definitely not obtained despite the identical signal-to-noise ratio. While the difference between the two estimates would probably be judged to be substantially relevant by most people, the experts in the field would have to assess whether the magnitude of the estimated pay gap itself (one-digit or very low two-digit area) qualifies as being economically relevant.

Taking up a setting where we are interested in a mean difference, the two-dimensional matrix of Table 5.3 describes four relevant cases that we might encounter when looking at an original study and a single replication study:

- **Case (1)** represents a situation where both studies obtain substantially similar (or even identical) effect size and standard error estimates. Necessarily, the signal-to-noise ratio z is also similar (or even identical). Irrespective of whether z is small or large, the two studies can be said to have produced significant sameness. When they are based on a small sample size, each study in isolation is left with a large remaining uncertainty (i.e., a large standard error and, therefore, a small z-ratio). However, even two small studies corroborate each other and the uncertainty is reduced when they find a substantially similar effect.
- **Case (2)** describes a situation where both studies obtain substantially similar (or even identical) estimates for the mean difference, but, largely due to different sample sizes, standard errors and signal-to-noise ratios are very dissimilar. The

replication context 1 from Table 5.2 is an example. As already discussed, obtaining a substantially similar (or even identical) effect size estimate in two studies is a mutual corroboration that represents "significant sameness"— irrespective of the signal-to-noise ratios of the studies.

- **Case (3)** describes two studies that obtain very dissimilar effect sizes but similar standard errors. Since the signal information is different, we cannot attach the label "significant sameness" to this research context.
- **Case (4)** describes a context where two studies obtain dissimilar effect sizes *and* standard errors. This includes the stylized replication context 2 from Table 5.2 where we happened to obtain an identical z-ratio because the much smaller effect size of the much larger second study was exactly offset by an equally smaller standard error. In other situations, such a computational offset might not apply and we might find differing z-ratios. Irrespective of z, studies with substantially different effect sizes cannot be said to have produced significant sameness.

When interpreting the constellations described in Table 5.3, we should avoid rash judgments regarding the quality of an original study when there is only one follow-up study. Especially in the case of small samples, substantially similar as well as substantially differing outcomes between the original and the replication study could be due to the vagaries of sampling.

To avoid confusion, we should also recall at this point that this book limits itself to inferential statistics that can be used as auxiliary tools for making inductive inferences (forming inductive beliefs). We are, therefore, also limited to considering random (sampling) error and inferential methods based on probability theory. Other types of error often cause problems in data-based research that impair inductive inference much more than the idiosyncrasies of random samples. Measurement problems and sample selection biases that cannot be corrected for are important examples (cf. Sect. 4.4). Such problems are not probabilistically accessible and can therefore not be dealt with by using inferential statistics. In other words, estimation uncertainty caused by random error is the only type of uncertainty the standard error and derived statistical quantities deal with (cf. McCloskey & Ziliak, 1996: 98). For example, in the replication context above, all arguments apply only if there are no errors besides random sampling error. Besides faultless measurements, this pre-supposes random sampling and an adequate consideration of the specific sampling design. Non-random errors, such as faulty measurements and uncorrected selection bias are beyond the reach of probability models and statistical inference and, therefore, also beyond the scope of this book.

5.3 Chapter Summary: Our Best Estimators Estimate Correctly on Average

The take-home message of this chapter is brief: in the most ideal case, we collected data through random sampling. If we then use adequate estimators, we can obtain an unbiased (random) sample-based effect size estimate (e.g., a mean difference or a regression slope) and a likewise (random) sample-based estimate of the uncertainty of this effect size estimation. Due to random sampling error, other studies based on random samples from the same population might obtain widely differing estimates for our population quantity of interest. However, unbiased estimators estimate correctly *on average* across many random samples. Therefore, it does not make sense to ask whether an individual study's sample-based estimate is "true" or not. Instead, we need to consider the body of evidence and include the weighted knowledge contribution of each properly made study even though, by its own, it might be too small to provide a reliable estimate. Otherwise, we will not obtain the most appropriate picture of the real-world effect. When the scales of the estimates from different studies are comparable or can be made comparable, meta-analysis can be used to synthesize the most plausible summary effect size that considers the accumulated evidence from a sequence of rigorous studies. Omitting a certain subset of studies from meta-analysis would inevitably produce a bias. For example, neglecting studies with low signal-to-noise ratios z when assessing the state of the knowledge in a certain field, as many researchers seem to be prone to do, would cause a systematic overestimation of the summary effect size.

In less ideal cases, we might only have a convenience sample. That is, instead of knowing the sampling design, we ignore how members of the population were selected into the sample. Sample statistics, which can of course be arithmetically produced, have therefore no inferential meaning: sample effect sizes such as mean differences or regression slopes do not represent unbiased estimates of the respective population effect sizes. The same applies to the sample statistic that, in a random sample, would be the standard error quantifying the uncertainty of the estimation. When the sample is not obtained through random sampling, it is impossible to use a sample quantity as estimate for random sampling error. In other words, non-probability samples preclude the use of inferential statistics because of assumptions violations. Consequently, applying meta-analysis to convenience samples would not make sense either because it is based on probability theory and the assumption of unbiased estimates that only vary due to random sampling error.

As already noted in Sect. 4.4, convenience samples are not without value and we can often make cautious inferences from observations in non-random samples. For example, we might have happened to observe an adverse health effect in a group of individuals who were haphazardly exposed to a certain substance. Even though the members of this group were not randomly selected and clearly non-representative of any target population of interest, it may be perfectly sensible to infer from these observations that other human beings are likely to suffer similar effects from this substance. This scientific inference would have to be based on medical reasoning that the fundamental physiological processes that caused the adverse health effect are similar if not identical in all people. However, random sampling being absent, this inference cannot be supported by inferential statistics. There are simply no sample statistics that inform us about the estimation uncertainty caused by random sampling error because there is no random sampling (cf. Fig. 2.1).

5.4 Recommended Reading

Andrews, I., & Kasy, M. (2019). Identification of and correction for publication bias. *American Economic Review, 109*(8), 2766–2794.

Becker, B. J., & Wu, M.-J. (2007). The synthesis of regression slopes in meta-analysis. *Statistical Science, 22*(3), 414–429.

Borenstein, M., Hedges, L. V., Higgins, J. P. T., & Rothstein, H. R. (2009). *Introduction to meta-analysis*. Wiley.

Card, N. A. (2012). *Applied meta-analysis for social science research*. Guilford.

Hirschauer, N., Grüner, S., Mußhoff, O., & Becker, C. (2018). Pitfalls of significance testing and *p*-value variability: An econometrics perspective. *Statistics Surveys, 12*, 136–172.

Deeks, J. J., Higgins, J. P. T., & Altman, D. G. (2021). Chapter 10: Analysing data and undertaking meta-analyses. In J. P. T. Higgins, J. Thomas, J. Chandler, M. Cumpston, T. Li, M. J. Page, & V. A. Welch (Eds.), *Cochrane handbook for systematic reviews of interventions version 6.2* (updated February 2021). www.training.cochrane.org/handbook

Chapter 6
The p-Value and Statistical Significance Testing

6.1 The p-Value Concept

To illustrate the ***p*-value concept**, we look at the sampling distribution of regression slope estimates as described in the upper part of Fig. 3.1. Let us recall the basics: in the case of a simple linear regression and a simple random sample (SRS), the sample slope b according to Eq. (3.5) can be used as unbiased estimate $\widehat{\beta}$ of the population slope β. The corresponding $\widehat{\text{SE}}$, i.e., the estimate of the true standard deviation SE of the sampling distribution, is obtained according to Eq. (3.6).

The upper part of Fig. 6.1 takes up the idea of the sampling distribution and is very similar to the upper part of Fig. 3.1. The only difference is that researchers now combine the estimated standard error with the so-called "point **null hypothesis**" (H_0) that the true regression slope in the population from which the random sample was drawn is zero ($\beta_{H_0} = 0$). With this assumption, the distribution of the estimates $\widehat{\beta}$ over frequent statistical replications would coincide with the distribution of the random sampling error $e_{\text{sam}} = \widehat{\beta} - \beta_{H_0}$ (cf. lower part of Fig. 3.1): under the null hypothesis, whatever effect would be observed in a particular random sample would be solely due to random error.

The lower part of Fig. 6.1 describes what happens when we divide the point estimate $\widehat{\beta}$ (signal) by its standard error estimate $\widehat{\text{SE}}$ (noise). When the sample size is "reasonably large" ($n \geq 30$; see Excursus), the signal-to-noise ratio $z = \widehat{\beta}/\widehat{\text{SE}}$ under the null hypothesis $\beta_{H_0} = 0$ can be assumed to have a standard normal distribution: $z \sim N(0; 1)$. Using the cumulative distribution function, we can now make statements about the probabilities of obtaining certain signal-to-noise ratios in frequently repeated random draws of equal-sized random samples—*conditional* on the null hypothesis of zero effect. We obtain, for example, a conditional tail probability $P(z \geq 1 \mid H_0) = 15.87\%$. Other familiar probabilities are $P(z \geq 1.645 \mid H_0) = 5\%$, $P(z \geq 1.96 \mid H_0) = 2.5\%$, and $P(z \geq 2.326 \mid H_0) = 1\%$.

© The Author(s), under exclusive license to Springer Nature Switzerland AG 2022
N. Hirschauer et al., *Fundamentals of Statistical Inference*, SpringerBriefs in Applied Statistics and Econometrics, https://doi.org/10.1007/978-3-030-99091-6_6

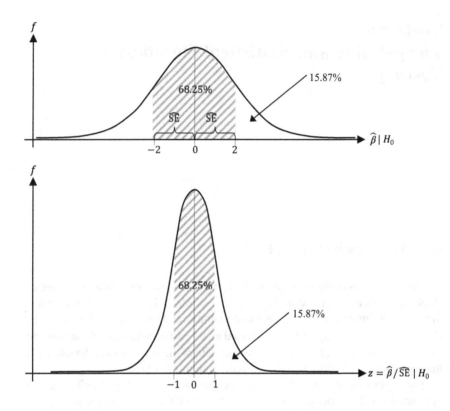

Fig. 6.1 Sampling distribution of a slope estimate $\hat{\beta}$ (upper part) and distribution of the z-ratio (lower part) conditional on the null hypothesis H_0: $\beta_{H_0} = 0$

If we made the simplifying assumption $\widehat{\text{SE}} = \text{SE}$, we could also make statements about the probabilities of obtaining certain estimates $\hat{\beta}$ in frequently repeated random draws of equal-sized random samples—*conditional* on the null hypothesis of zero effect. For example, if we assumed $\widehat{\text{SE}} = 2 = \text{SE}$, we would obtain a tail probability $P\left(\hat{\beta} \geq 2 \mid H_0\right) = 15.87\%$ (cf. upper part of Fig. 6.1). Since we are interested in the uncertainty of the estimation, this perspective seems natural. However, we cannot make conclusive statements regarding the tail probabilities of point estimates *conditional* on the null hypothesis because the standard error is itself but an estimate. With both the numerator (the point estimate $\hat{\beta}$; signal) and the denominator (the estimated $\widehat{\text{SE}}$; noise) being subject to sample-to-sample variation, a certain signal-to-noise ratio z can result from different point estimates.

Excursus: To reduce the noise caused by the vagaries of random sampling, researchers should generally try to collect "reasonably large" samples. We assume throughout this book that reasonably informative data collection approaches with sample sizes $n \geq 30$ are assured.

Following the rule of thumb "$n \geq 30$" generally enables researchers to use two convenient approximations based on the central limit theorem: first, we can base inferential reasoning on the assumption that the sampling distribution of the point estimate (signal) is approximately normal under regular conditions (cf. Sect. 3.2, Footnote 3). Second, we can approximate the distribution of the signal-to-noise ratio through a normal distribution with a standard deviation of 1. This holds even though the standard error (noise), which is based on the sum of squares, exhibits a χ^2-distribution (chi-squared). While the quotient of a normally distributed signal (numerator) and a chi-squared noise (denominator) exhibits a t-distribution with $n - 1$ degrees of freedom, it is asymptotically normally distributed and can therefore be approximated through a normal distribution in sufficiently large samples.

Indicating the respective distributional assumptions, the signal-to-noise ratio under the null hypothesis is called a t-ratio ("t-statistic")—instead of a z-ratio ("z-statistic")—when we do not approximate the t-distribution by the standard normal distribution ("z-distribution"). For example, when we are interested in a mean difference [cf. Eqs. (3.3) and (3.4)], the label "*two-sample-t-statistic*" is commonly attached to the signal-to-noise ratio $\widehat{\delta}/\widehat{SE}_d$ obtained from the sample. In sufficiently large samples, we could resort to the normal approximation. If we do so, we attach the label "*two-sample-z-statistic*" to the same sample-derived value of $\widehat{\delta}/\widehat{SE}_d$.

For $n \geq 30$, the differences in the tail probabilities between the exact and the asymptotic distribution are minor. For example, in the "asymptotic" standard normal distribution, we have a tail probability $P(z \geq 1.645) = 0.050$. With sample size $n = 30$, the tail probability according to the "exact" t-distribution with 29 degrees of freedom is $P(t \geq 1.645) = 0.055$.

Following a still widespread statistical routine, many researchers near-automatically aggregate an effect size estimate such as $\widehat{\beta}$ and its standard error $\widehat{SE}$ into a signal-to-noise ratio $z = \widehat{\beta}/\widehat{SE}$ ("z-statistic"). They then report the estimate $\widehat{\beta}$ together with a so-called **p-value**, which is based on the null hypothesis of zero effect. Figure 6.2 (upper part) illustrates that the p-value is simply the label attached to the **conditional tail probability** for a particular z-ratio: $p\left(z = \widehat{\beta}/\widehat{SE}\right) = P\left(z \geq \widehat{\beta}/\widehat{SE} \mid H_0\right)$. For example, after having obtained a slope estimate of $\widehat{\beta} = 3.29$ and a standard error estimate of $\widehat{SE} = 2$, which results in $z = 1.645$, the conventional routine of presenting results would be to report the slope estimate of 3.29 together with a p-value of 5% ($= P(z \geq 1.645 \mid H_0)$). Unfortunately, this routine is

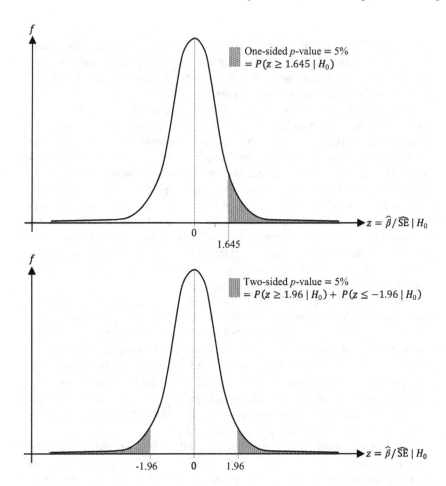

Fig. 6.2 One-sided and two-sided p-value of 5% derived from the distribution of the signal-to-noise ratio $z = \widehat{\beta}/\widehat{SE}$ under the null hypothesis H_0: $\beta_{H_0} = 0$

usually performed without explaining why it should be useful to compare the data with a postulated effect of zero in the specific research context.

Instead of taking a one-sided look at either one of the tails of the z-distribution under the null hypothesis, researchers usually report a two-sided p-value when they have no prior assumption regarding the direction of an effect. The lower part of Fig. 6.2 shows that $z = 1.96$, as resulting, for example, from a slope estimate $\widehat{\beta} = 3.92$ and a standard error estimate $\widehat{SE} = 2$, would be needed to obtain a two-sided p-value of 5% ($= P(z \geq 1.96 \mid H_0) + P(z \leq -1.96 \mid H_0)$).

So how can we intelligibly express the **meaning of a p-value**? Popular but inaccurate explanations similar to the following are often repeated: "The p-value is the *conditional* probability of re-finding the *observed* estimate or a larger one in *frequently* repeated random draws of equal-sized samples *if* we assumed the null hypothesis of zero effect to be true in the parent population." For this definition to be correct, we would have to assume that the sample-based estimate of the standard error $\widehat{SE}$ happens to coincide with the true standard deviation SE of the sampling distribution. When we correctly account for the sample-to-sample variability of standard error estimates, we have to restrict the *p*-value definition to the signal-to-noise ratio. While indeed sounding complex, the correct definition for a one-sided *p*-value reads as follows: **"The *p*-value is the *conditional* probability of re-finding the *observed* signal-to-noise ratio or a larger one in *frequently* repeated random draws of equal-sized samples from a population *if* we assumed the null hypothesis of zero effect to be true in that population."** In a very abridged way, we could also say that the *p*-value indicates the compatibility of the data with a specified statistical model including the null hypothesis of zero effect.

So what can we learn from the *p*-value? The original evidence that can be derived from a random sample is, at best, an unbiased point estimate (signal) and an unbiased estimation of the uncertainty of this point estimation (noise) expressed through the standard error; no more, no less. This information is partly lost and what remains as information becomes opaque when signal and noise are aggregated into a *z*-ratio or even a *p*-value. Serious cognitive failures are imminent when the mathematical manipulations underlying the *p*-value are only superficially remembered but not vividly present in researchers' minds.[1] To avoid misinterpretations of the *p*-value, one should therefore always keep in mind the implications that inevitably result from its mathematical construction:

- Being a *conditional* tail probability based on the *assumption* that the null hypothesis of no effect is true in the parent population, the *p*-value is *not* the probability that the null hypothesis is true. The *p*-value is therefore *not* the probability either of making an error when rejecting the null hypothesis.
- Any non-zero point estimate leads to a two-sided *p*-value of less than 1 (and a one-sided *p*-value of less than 0.5). In other words, obtaining a *p*-value of less than 1 (or less than 0.5 in a one-sided perspective) implies that the null hypothesis is *not* the proposition the data are most compatible with. We might also simply say that the effect size observed in a random sample is always the estimate that is best supported by the data, regardless of its *p*-value.

---

[1]The same applies to confidence intervals (CIs). While the term appears to confuse even experienced statistical practitioners, a CI is just an alternative way of reporting point and standard error estimates derived from a single random sample. For example, a point estimate $\widehat{\beta} = 1$ together with a standard error estimate $\widehat{SE} = 1$ could be reported as a 95% CI $[\widehat{\beta} - 1.96 \cdot \widehat{SE}, \ \widehat{\beta} + 1.96 \cdot \widehat{SE}]$ or simply $[-0.96, \ 2.96]$. A popular misinterpretation of such a result is be to believe that the true population parameter is within this interval with a probability of 95%. This is and must be wrong for the reason alone that, due to random sampling error, we would find different point and standard error estimates (and thus different CIs) in different random samples.

- As illustrated through Eqs. (3.2), (3.4), and (3.6), the estimated standard error $\widehat{SE}$ for point estimates such as a mean, a mean difference or a regression slope becomes the smaller the larger the random sample. And being a quotient of the point estimate (signal) and its estimated standard error (noise), the z-ratio becomes the larger the smaller the standard error.
- Being the tail probability of the z-ratio under the null hypothesis, the p-value, therefore, necessarily becomes small when sample size is large. Thus, even a population effect size that is clearly too small as to be substantially (e.g., economically) relevant will produce small p-values in large samples.
- The p-value is a non-metric statistic: a difference between, say, a p-value of 0.30 and 0.29 does not indicate the same increase of the strength of evidence (exceedance of the signal over the noise) as a difference between 0.05 and 0.04.

6.2 Null-Hypothesis-Significance-Testing

6.2.1 Dichotomization of the p-Value and Significance Declarations

The statistical practice still in dominant use in the social sciences is based on the **"null-hypothesis-significance testing"** (NHST) framework, which is also often briefly spoken of as (statistical) **"significance testing"** or **"hypothesis testing"** (see Sect. 6.3 why both terms are confusing misnomers for NHST). NHST is based on an **arbitrary dichotomization** of the signal-to-noise ratio and, therefore, the p-value. A z-ratio of 1.645 (one-sided perspective) or 1.96 (two-sided perspective) and the corresponding p-value of 5% are widely used as thresholds. Researchers who use the NHST-approach and the conventional threshold of 5% often use inadequate wordings similar to the following to describe their sample-based estimates:

- They attach the label *"statistically significant"* (or simply, "significant") to estimates associated with $p \leq 0.05$ and earmark such estimates with an asterisk (*). They might also say that they *"can reject the null hypothesis at the significance level of 5%."* In a very abbreviated form, they might call their sample-based estimate a *"positive result."*
- Analogously, they attach the label *"statistically non-significant"* (or simply, "non-significant") to estimates associated with $p > 0.05$. They might also say that they *"cannot reject the null hypothesis at the significance level of 5%"* or that they found a *"negative result."*

Many researchers perform this procedure much like an automatic routine. That is, they neither provide a context-specific justification for using a particular p-value threshold such as 0.05 nor an explanation of why they think it is useful to assess the compatibility of the data with a hypothetical effect of zero. Therefore, it is difficult to understand what researchers want to communicate when they claim that the null hypothesis can or cannot be rejected at the significance level of 5%. But even if one accepts the convention of making dichotomous statements, there are variations of wordings that further increase confusion and cause interpretations that are wrong even within the NHST-framework. It is wrong, for example, to interpret a non-zero effect associated with $p > 0.05$ as a *confirmation of the null hypothesis*—instead of noting that the *null hypothesis cannot be rejected* at the conventional significance level of 5%. Not having obtained a signal-to-noise ratio that is large enough to reject the null hypothesis according to NHST does not represent "evidence of absence." It would even be logically inconsistent to interpret an effect associated with $p > 0.05$ as "absence of evidence." The evidence in terms of the effect size (signal) observed in the sample is just not strong enough in relation to the standard error (noise) to reject the null hypothesis according to NHST-convention.

At first view, it may seem convenient to resort to a simplified representation of results and report point estimates accompanied by statistical significance declarations. We must realize, however, that the inferential task is to assess the validity of sample-based estimates ("make sense of our figures"). That is, we must make **reasonable inferences** from the sample toward its parent population (inferential target population) given prior knowledge (e.g., from previous studies) *and* the incremental information extracted from the particular sample. While at the very best we can reduce uncertainty (through larger samples) and assess uncertainty (by correctly estimating the standard error), we must embrace the remaining uncertainty caused by random sampling error as long as we study samples instead of full populations. An oversimplified representation of results does not help us answer the question of what we should most reasonably believe after seeing the results of a study. On the contrary. Downgrading the intelligible information regarding a point estimate and its uncertainty into a binary variable of the form "significant" vs. "non-significant" propagates cognitive biases. It virtually seduces researchers and users of research results alike to draw overconfident conclusions, both when p is below and when it is above the arbitrary significance threshold.

We can conclude that even in the best case when there are no misinterpretations of statistical significance declarations, a large part of the original information that can be derived from a random sample is lost when the point estimate (signal) and its standard error estimate (noise) are aggregated into a z-ratio or a p-value, or even squeezed into a dichotomous significance declaration. This is due to three reasons:

1. Very different combinations of signal and noise can lead to identical z-ratios. We would have to look back at the numerator (point estimate) and the denominator (standard error) anyhow to access the crucial information about the size of the estimate and the random-error-related uncertainty of that estimation.
2. The null hypothesis of zero effect that underlies the p-value concept is usually an uninteresting hypothesis to challenge and contrast data with. This is because we often know *a priori* that an exact zero is not true. For example, it is unlikely that

any two variables exhibit exactly zero correlation. Hence, reporting the estimate's size (distance from zero) and uncertainty is more informative.[2]

3. As with any downgrade of the measurement scale, even more information is lost when continuous p-values are dichotomized by using arbitrary significance thresholds such as 0.05.

6.2.2 The Statistical Ritual "NHST" and Misinterpretations of Single Studies

There has been long-standing criticism that the NHST-routine almost inevitably causes inferential errors. It makes people forget that what we have, at best, is an unbiased point estimate and an unbiased estimate of the uncertainty of this estimation derived from a random sample. Some critics describe NHST-practices even as "**statistical rituals**" based on collective delusions that have largely eliminated critical thinking in the social sciences (e.g., Gigerenzer, 2004/2018; Ziliak & McCloskey, 2008). These delusions make researchers believe that statistical significance declarations facilitate automatic scientific inferences. The fact that a vast body of literature has decried misinterpretations of "statistical significance" for more than half a century (cf. Hirschauer et al., 2016/2018 for an overview) has apparently been of little avail. The ubiquity and persistence of faulty interpretations and practices are, not least, caused by the fact that they have been perpetuated for decades through inadequate teaching and even best-selling statistics textbooks that usually focus strongly on NHST and on achieving "statistical significance."[3]

While misinterpretations have been criticized for a long time, the scientific debate has intensified from 2010 onward and there have been prominent calls to abandon NHST. In 2016, the *American Statistical Association* (ASA) issued an unprecedented methodological warning that stressed that p-values and statistical significance can neither be used to determine whether a hypothesis is true nor whether a finding is

---

[2]Imbens (2021) agrees that a *point estimate* and its *uncertainty* rather than a p-value are the relevant output from statistical analysis in most economic research settings. But he sees a role for p-values *if* substantial prior probability can be put on the null. That is, rather than reflecting an unlikely proposition, the null must be specified to reflect the most plausible prior scientific belief. Imbens argues that, in this case, a low p-value and, therefore, a low compatibility of the data with the established scientific belief is informative. He cautions, however, that a high incompatibility of a single dataset with a strong prior scientific belief can only serve as an indication that it might be worthwhile investigating the issue further with new data (*"worth a second look"*). While providing a meaningful interpretation of p-values, Imbens' argument leaves open the question of why p-values should better indicate what is worth a second look than *separately reported effect sizes and standard errors*. Effect sizes that are only marginally different from zero are associated with small standard errors and p-values in large samples, but they might not be worth a second look.

[3]For references to faulty textbooks and textbooks that have perpetuated inferential errors due to their strong focus on significance testing procedures and their little emphasis on the understanding of statistical results, see Nickerson (2000), Haller and Krauss (2002), Ziliak and McCloskey (2008), Krämer (2011), Lecoutre and Poitevineau (2014), and Gigerenzer (2018).

important (Wasserstein & Lazar, 2016). Three years later, *The American Statistician* published a special issue "Statistical Inference in the 21st Century: A World Beyond $p < 0.05$." Summing up the reform suggestions, the editors state that it is time to **abandon statistical significance testing** (Wasserstein et al., 2019). Almost simultaneously, a widely supported call to "Retire statistical significance" was published in *Nature* (Amrhein et al., 2019). In the same year, the National Academies of Sciences (2019) took up the criticisms in its "Consensus Report on Reproducibility and Replicability in Science."

Despite well-founded criticisms, misinterpretations of NHST continue to be an alarmingly "normal" practice in many fields. While there are noteworthy early exceptions, including the publications by McCloskey and Ziliak (1996), Ziliak and McCloskey (2008), or Krämer (2011), the acknowledgment of the problems associated with NHST seems to lag behind and be weaker in economics at large compared to many other fields. Therefore, scientists concerned with quantitative research in economics are still likely to encounter statistical significance declarations and overconfident yes/no conclusions in research papers, dissertations, statistics textbooks, and statistics teaching. Aiming to provide an understanding of why these errors occur, we describe statistical practices associated with the NHST-routine in regression analysis, which is the working horse of many studies in the social sciences. The most common misinterpretations associated with the NHST-approach are best understood when realizing that many regression analyses are performed as if using the following, **largely misleading working steps**:

1. Put much emphasis on model specification and run a multiple regression to identify regression slope coefficients ("effect sizes"). Call the identification of sample regression slopes "estimation" without reflecting whether they can be used as pertinent estimates for the unknown population slopes.
2. Do not articulate from which parent population your sample was drawn or whether you have a random sample in the first place, but automatically report statistics labeled as estimates, standard errors, and p-values.
3. Automatically attach the label "statistically significant" or "positive result" to slope estimates with p-values below a threshold (usually 0.05). Analogously, use the label "statistically non-significant" or "negative result" for estimates associated with $p > 0.05$. Always follow this procedure without reflecting on the usefulness of the null hypothesis of zero effect and the arbitrary dichotomization of results into "significant" and "non-significant" at the given threshold.
4. Implicitly attach a high trustworthiness or probability to "significant" slope estimates (i.e., ones that are associated with $p \leq 0.05$) or even claim the estimated effect to be real and replicable. Do not discuss the effect size or even insinuate that a "significant" effect is large or important. Analogously, attach a low trustworthiness or probability to "non-significant" slope estimates (i.e., ones associated with $p > 0.05$) or even claim that there is no (noteworthy) effect.

So what is correct and what is faulty with this NHST-procedure that seems to be the standard routine in many studies?

Assessment of Step 1 It is crucial to identify a regression model that adequately considers the data structure, irrespective of whether we have a random sample or not. For example, modeling a linear association between two variables X and Y does not adequately describe their relationship when, in the data, this relationship is exponential rather than linear. While being common usage, it should be noted that attaching the label "estimation" to the identification of an adequate sample regression slope b might be confusing. This is because, similar to simpler sample quantities such as the mean $\bar{x}$ or the mean difference d, a sample slope is first of all a summary description of a property of the *given* dataset. Whether the X-Y-relationship identified in the data can be used as an *estimate* for the unknown X-Y-relationship in the target population is a separate issue (cf. step 2).

Assessment of Step 2 A virtually unlimited number of summary statistics (sample quantities) can be arithmetically computed for any set of data. But quantities that would carry inferential meaning in the case of random sampling are devoid of such meaning when there was no probabilistic data generation process. For example, claiming that a computed quantity is a standard error (i.e., the estimated standard deviation of the sampling distribution) or a p-value (i.e., the tail probability of the signal-to-noise ratio under the null hypothesis) is wrong when the data that enter the computation are not from a random sample. Reporting a "standard error" in such circumstances would deceptively suggest that we can assess random sampling error even though there is no probabilistic model to do so. When we only have a convenience sample but not a random sample, we should not report quantities that could be mistaken to carry inferential meaning (cf. Sect. 4.4).

Assessment of Step 3 Downgrading the continuous signal and noise information (i.e., the point and the standard error estimate) into a significance declaration (i.e., a dichotomous variable based on an unquestioned null hypothesis of zero effect and an arbitrary threshold) is not by itself faulty but causes a substantial loss of information and is deceptive. One should be able to justify such practice by explaining how inference is better supported by dichotomous significance declarations than by reporting the "original" evidence in terms of the estimated effect size and the uncertainty of this estimation. In most research contexts (if not all) where we are interested in forming the most reasonable inductive belief regarding a real-world state of interest, no such justification will be found. Therefore, the habit of routinely making dichotomous significance declarations should be abandoned.

Assessment of Step 4 Dichotomous significance declarations make not only lay people but also many researchers draw dichotomous existence or relevance conclusions (yes/no conclusions) that—to borrow a quote from H.L. Mencken in the New York Evening Mail from November 16, 1917—are "neat, plausible, and wrong." The first neat but wrong conclusion is that effects with the label "statistically significant" can be considered to be real or replicable with a high probability. The second neat but wrong conclusion is that the label "statistically non-significant" is an indication or even proof of no or little effect. Interpretations along this erroneous dichotomy are entrenched practice for many researchers, a practice that is based on deeply ingrained delusions that result both from wishful thinking (desire for a "neat interpretation") and the inevitably wrong connotation of the word "significance" in everyday language.

We may summarize that the *p*-value and statistical significance have much less inferential content than what colloquial associations with the terms "statistically significant" and "statistically non-significant" suggest. Following the common routines described above mirrors serious misunderstandings of what information the *p*-value can provide. In the decade-long scientific debate on inferential errors associated with NHST, the most widespread and devastating inferential errors have been given prominent names:

Inverse Probability Error   Cohen (1994: 997) used this term to describe the belief that the *p*-value is the probability of (falsely rejecting) the null hypothesis given the data. He succinctly described the harmful mix of wishful thinking and semantic confusion that causes this delusion: "[the *p*-value] does not tell us what we want to know, and we so much want to know what we want to know that, out of desperation, we nevertheless believe that it does! What we want to know is 'given these data, what is the probability that H0 is true?' But [. . .], what it tells us is 'given that H0 is true, what is the probability of these (or more extreme) data?' [. . .]." Contrary to the delusive NHST-wordings that speak of "hypothesis *testing*" and "*confirmatory* analysis," a *p*-value cannot be used to "test" or "confirm" a hypothesis in any usual sense of the word. A *p*-value describes data frequencies *assuming* a particular statistical model including the null hypothesis. Therefore, it cannot work inversely and tell us whether a hypothesis is true or what is its probability.

Sizeless Stare   Ziliak and McCloskey (2008) coined this expression for the disregard of effect size or the implicit but erroneous equation of statistical significance with relevance. In the light of the increasing availability of large samples, the naïve equation of significance with relevance becomes more and more misleading because even the most marginal effects become "statistically significant" in large samples. This has serious implications because it is unlikely that any two variables exhibit zero correlation. This is why the null hypothesis is sometimes called a "straw man" that significance testing tries to knock down. Already back in the 1970s, Leamer (1978: 89) put this in a nutshell by noting that "since a large sample is presumably more informative than a small one, and since it is apparently the case that we will reject the null hypothesis in a sufficiently large sample, we might as well begin by rejecting the hypothesis and not sample at all."

Replication Fallacy   Gigerenzer (2004/2018) used this label to describe the erroneous belief that, in the case of statistical significance, $1 - p$ indicates the probability that significance will be re-found in other equal-sized random samples from the same parent population (i.e., in "exact statistical replications"). He argues that this delusion has been particularly devastating for the social sciences because it made researchers believe that there is no need to carry out replications to scrutinize studies whose results were "significant." Accordingly, he claims that the ritual "NHST" has deformed disciplinary cultures, which are now overly focused on single studies with "significant" results, and which, contrary to the natural science, put too little emphasis on replication as a necessary and valuable principle of good science in a world fraught with uncertainty.

6.2.3 Perpetuation of the Statistical Ritual "NHST" in Replication Studies

In recent years, the need for **replication studies** has been increasingly acknowledged in the social sciences and large replication efforts have been undertaken (e.g., Camerer et al., 2016/2018; Dreber & Johannesson, 2019). While taking up some criticisms regarding misapplications associated with NHST, the problem of many replication studies is that they are still stuck in a dichotomous classification of results as being either "positive" (significant) or "negative" (non-significant). That is, the assessment of what is a replication success or failure remains itself based on statistical significance. For example, a study with a "significant" finding followed by a study with a "non-significant" finding is often perceived as replication failure. This is a misleading interpretation for several reasons:

1. Using a fixed significance threshold is arbitrary: there is neither a justification for a standard level of 0.05 nor for a level of 0.04 (or any other). Therefore, a sharp classification of results as being "positive" or "negative" depending on, say, a p-value of 0.049 as opposed to 0.051 is clearly not helpful for assessing whether a replication was successful or not.
2. Even estimates with *very* different signal-to-noise ratios z and, thus, *very* different p-values are not necessarily conflicting. Since the z-ratio is a quotient, a very similar or even identical point estimate obtained in the follow-up study can be associated with a very dissimilar z-ratio and p-value—especially when different sample sizes are used. The opposite also applies: even very dissimilar estimates can be associated with very similar or even identical z-ratios and p-values (cf. Tables 5.2 and 5.3).
3. Unbiased estimators estimate correctly on average (cf. Table 5.1). Therefore, even estimates and p-values from studies with identical sample sizes may exhibit considerable sample-to-sample variation. We consequently need the estimates from all properly implemented studies, irrespective of their p-values, to obtain the most appropriate picture of the population effect size.

Using statistical significance as replicability criterion is a fallacy similar to "vote counting" in narrative reviews that contrast the tallies of "significant" and "non-significant" findings to assess the state of knowledge in a certain research field. Borenstein et al. (2009: 14) note that doing narrative reviews based on vote counting boils down to "doing arithmetic with words" and that "when the words are based on p-values the words are the wrong words." They claim that this problem "gallops"

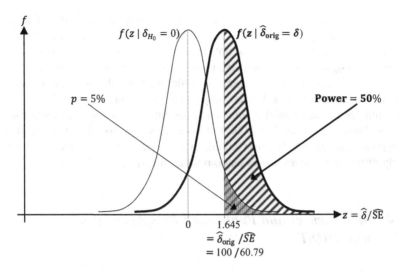

Fig. 6.3 The distribution of the signal-to-noise ratio z under the null hypothesis $\delta_{H_0} = 0$ and under the true mean difference $\delta = 100$

through many research fields. This criticism is directly related to the interpretation of replications based on the statistical significance criterion.

We resort to the gender pay gap example in Table 3.1 to illustrate why statistical significance is an inadequate replicability criterion. Study 1, which is taken to be the original study, finds a mean difference $\widehat{\delta}_{orig} = 100$ in favor of men in a random sample of 300 men and 300 women, with standard deviations sd $= 744.5$ in both groups. This leads to an estimated standard error $\widehat{SE} = 60.79$, a signal-to-noise ratio $z = 1.645 = 100/60.79$, and a one-sided $p = 0.05$. Assume now that this estimate exactly matches the true gender pay gap in Phantasia $(\widehat{\delta}_{orig} = \delta)$. In this case, equal-sized replication studies will only have a "power" of 50% (cf. Fig. 6.3).

What does "power" mean? **Power** is the estimated probability of obtaining statistical significance over frequent statistical replications given the true population effect size. Figure 6.3 illustrates the concept for the gender pay gap example: with the true mean difference being $\delta = 100$, the sampling distribution can be considered to be approximately normal according to the central limit theorem: $\widehat{\delta} \sim N(\mu; \sigma^2)$, with $\mu = 100$ and $\sigma = \widehat{SE} = 60.79$. The corresponding distribution of the signal-to-noise ratio is: $z \sim N(1.645; 1)$. That is, we are likely to obtain signal-to-noise ratios $z \geq 1.645 = 100/60.79$ (and, therefore, $p \leq 0.05$) in only 50% of equal-sized replication studies.

Even though the original study happened to find exactly the true gender pay gap, only 50% of equally large follow-up studies will successfully "replicate" in terms of obtaining $z \geq 1.645$ and, therefore, differences that would be called "statistically significant" $(p \leq 0.05)$. This low "replicability" would probably be considered as disastrous by many people who are interested in assessing the quality of prior research. However, it is not an indication of bad science but the mathematical implication of using the statistical significance criterion for judging whether a finding is successfully replicated or not. Instead of using a misleading classification into positive (significant) and negative (non-significant) results, a reasonable evaluation of a replication study requires considering both the uncertainty and the magnitude of its effect size estimate compared to the original study. When the effect size estimates are substantially similar (consistent), i.e., when "**significant sameness**" is obtained (cf. Sect. 5.2), both studies can be said to corroborate each other. In the opposite case, it must be observed that even perfectly rigorous studies can obtain widely differing estimates due to random sampling error.

6.2.4   Malpractices and Publication Bias Associated with NHST

Besides misinterpretations of *p*-values and statistical significance tests, improper practices associated with the statistical routine of NHST cause further inferential errors. The harmful consequences of these malpractices are generally known as **publication bias** (Smith, 1980). Publication bias describes the problem that the body of evidence that is made available to the scientific public is distorted toward results that can be declared "statistically significant." This is due to ill-guided incentives and the present scientific publishing system that prefers statistically significant results. While Sterling (1959) pointed out the problem already in the 1950s, it is still highly relevant. In the 1970s, Rosenthal (1979) coined the metaphor "file drawer problem" to describe the mechanism that underlies publication bias. According to this view, journals are filled with the 5% of the studies that happened to find statistically significant results ("positive results"), while researchers' file drawers are filled with the 95% of the studies that never saw the light of day because the findings were not statistically significant ("negative results").

While some researchers may honestly but erroneously believe that "starless" results are not interesting enough to warrant publication, the major problem is highlighted by the publish-or-perish witticism. In our competitive research system, most researchers are under pressure to produce papers with novel findings. If papers with statistically significant findings are more likely to be published, researchers are likely to adopt harmful selection strategies:

- **Selective preparation** refers to researchers' decisions not to conduct studies in the first place that are likely to be a "waste of time and money" because they do not promise to produce "significant results."
- **Selective submission** means that, anticipating low chances of being published, researchers do not submit the results of existing studies to journals when those studies happened to produce "non-significant results."

- **Selective reporting** describes researchers' self-interested but not scientifically justified selection of one of many analytical variants that "worked best" in terms of producing significance.

Closely related with researchers' selective choices, there is the problem of **selective publishing decisions** due to reviewers' and editors' preferences for statistical significant results. On the one hand, selective publishing decisions fuel researchers' harmful selection strategies because they are anticipated. On the other hand, they are a cause of bias in their own right because studies with "non-significant results" that are submitted by researchers who are conscientious enough to refrain from self-interested selection practices have little chances of being published.

As a result of all these selection choices, "significant findings" are overrepresented and studies with "non-significant results" tend to never being presented to the public. This leads to the definition by Kline (2013: 274) according to which reporting and publication bias implies that published studies have more statistically significant findings and larger effect sizes than unpublished studies. Sections 3.3.3 and 5.1.2 have already shown that the latter necessarily results from focusing on statistically significant findings because it is the "abnormally" large sample effect sizes that produce very high signal-to-noise ratios (Trafimow et al., 2018). Consequently, an upward bias arises if we consider only the results of studies with high signal-to-noise ratios (e.g., those above a certain threshold such as $z = 1.645$ or 1.96)—instead of using all estimates, wherever they are located in the sampling distribution, as is required for an unbiased assessment of the population effect size.

Over the last decades, a variety of meta-analytical methods have been developed to gauge publication bias (for an overview see Rothstein et al., 2005; Cooper et al., 2009; Weiß & Wagner, 2011; or Hirschauer et al., 2018). In many fields, publication bias itself has become an object of study and a large number of studies have found substantial publication bias across many disciplines (e.g., Gerber et al., 2010; Auspurg & Hinz, 2011; Joober et al., 2012; Head et al., 2015; Berning & Weiß, 2016; Brodeur et al., 2016/2020; Isaiah & Kasy, 2019). The file drawer effect leads not only to a situation where scientific claims are too rarely subjected to independent scrutiny because researchers shy away from replication efforts that are likely to produce "negative results." It may also lead to wrong decisions and policies with dire practical consequences because the scientific community does not succeed in making available all evidence that is needed to arrive at the most reasonable belief regarding a real-world state of interest.

Any selection strategy—whether it is selective preparation, submission, reporting, or publishing—that is aimed at achieving statistical significance instead of knowledge causes knowledge distortions. While it would not be easy to make a general statement which of the above-described selection strategies is most harmful, selective reporting has become the most widely criticized one in recent years. Commonly, the term "**p-hacking**" is used to describe selective reporting practices motivated by researchers' desire to obtain statistically significant (and presumably

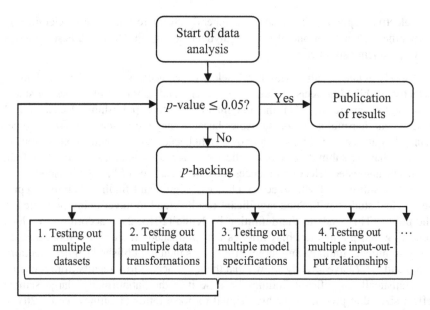

Fig. 6.4 The manifold possibilities of *p*-hacking. Source: Own representation based on Motulsky (2014: Fig. 1).

publishable) results.[4] One might speculate that the critical debate focuses on *p*-hacking because omissions to present and publish all results are considered as less serious infringements of good scientific practice than outright manipulative actions aimed at producing statistical significance.

The literature concerned with the pitfalls of significance tests and declarations has extensively discussed various forms of *p*-hacking (see Hirschauer et al., 2016/2018 for an overview). Figure 6.4 provides a brief systematic overview to facilitate the understanding in which ways *p*-hacking represents a covert and thence especially harmful type of multiple testing.

1. **Covertly testing out multiple datasets:** Researchers might be tempted to explore whether *p*-values can be reduced when the number of units in a sample is manipulated. There are several possibilities: (i) A first *p*-hacking opportunity is the post-design increase of sample size if an original sample yielded "disappointing" *p*-values. A general feeling that larger samples are better may impede the awareness that this represents a noxious case of selective reporting after testing

---

[4]The term "*p*-hacking" is ascribed to Simmons et al. (2011: 1359) who state: "[...] it is unacceptably easy to publish 'statistically significant' evidence consistent with *any* hypothesis. The culprit is a construct we refer to as *researcher degrees of freedom*. In the course of collecting and analyzing data, researchers have many decisions to make: Should more data be collected? Should some observations be excluded? [...] it is common (and accepted practice) for researchers to explore various analytic alternatives, to search for a combination that yields 'statistical significance', and to then report only what 'worked'."

out multiple sample sizes. This is because the decision to collect more data is only taken if the original sample failed to produce low *p*-values. (ii) Another one is to explore how *p*-values can be reduced by removing more or less data from the analysis. Examples are the elimination of particular data points on the seemingly justifiable grounds of being outliers, or the use of a "promising" data subset for the analysis. (iii) A related *p*-hacking opportunity is to systematically test out which *p*-values can be obtained when analyzing multiple data subsets. Imagine a group of researchers is interested in the relationship between body height and per capita income. Having a large sample with, say, $N = 4000$ subjects, they subdivide it into 20 subgroups of 200 subjects and then analyze each subgroup separately. Even when there is no association between body height and income in the data, they are quite likely to find a statistically significant link between body height and income in at least 1 of the 20 subgroups. Now imagine the researchers suppress the results of the other subgroups and selectively report this "statistically significant result" under the heading "Strong evidence for relationship between body height and income." While this would obviously be a serious case of self-serving *p*-hacking, it should be noted that it actually produces the same detrimental distortion as selective publication decisions resulting from editors' preferences for "positive results." To understand this, imagine 20 research groups tackle the research questions independently based on small samples of 200 subjects each. Similar to the blatant *p*-hacking blunder described above, the body of evidence would be distorted because only a study that happened to obtain a statistically significant relationship would see the light of the day.

2. **Covertly testing out multiple data transformations:** Researchers might also be tempted to check which (combination) of the many conceivable data transformations produces lower *p*-values than the original data. Possibilities are plentiful: (i) Downgrading of measurement scales (e.g., age classes instead of age in years), (ii) log-transformations or squaring, and (iii) the synthetization of various variables, including ratios (e.g., risk ratios, odds ratios, etc.) and interaction terms. Some of these manipulations may be appropriate in the light of the theory, research question, and data. However, we are facing an overestimation of the validity of the empirical evidence if these data manipulations are driven by significance-pursuing behavior and combined with selective reporting practices.

3. **Covertly testing out multiple model specifications:** In regression analysis, there is ample scope for *p*-hacking even if the original sample size and data are left unaltered. Again, there are several possibilities. (i) A first one arises because the choice of variables to be included in the model is often ambiguous. This applies to control variables as well as to the variables that are used to measure the response variable and the focal explanatory variables. Significance-pursuing researchers might be tempted to mine for a variable combination that yields low *p*-values. Imagine researchers want to study how people's attitudes toward organic farming are related to their willingness to pay for organic products. Imagine also that information on the latent variable "attitude" is collected via several Likert-scaled survey items (= manifest variables). We undoubtedly face a distortion and an overestimation of the empirical evidence if the researchers covertly mine for an

item or index for "attitude" until they find one that produces a "significant" result. (ii) Beyond the manipulative choice of the variable set, the ambiguities in the selection of the structure of econometric models also offer ample scope for self-serving decisions that yield "statistical significance." Imagine we are facing the choice of whether to use a pooled regression model or rather a panel data model. We often have to make such decisions to the best of our knowledge, and it is often scientifically advisable to triangulate models and openly compare the results of different model specifications. However, the rules of good scientific practice are broken when the data analysis is not performed as planned in a prior study design but ad hoc adjusted according to the criterion of which analytical model yields low *p*-values. Transparency, *the* prerequisite of meaningful scientific communication, is completely lost when the results of competing models are neither explicitly reported nor comparatively discussed. In brief, just as other significance-pursuing behaviors, *p*-hacking regarding the specification of estimation models introduces bias.

4. **Covertly testing multiple input–output relationships:** The most devastating *p*-hacking practice is probably when researchers covertly subject multiple input–output relationships in *one* dataset to significance testing. Kerr (1998) coined the term HARKing—**H**ypothesizing **a**fter the **R**esult is **K**nown—to point out the essential characteristic of this malpractice. Motulsky (2014: 201) further explains: "This is when you analyze the data many different ways (say different subgroups), discover an intriguing relationship, and then publish the data so it appears that the hypothesis was stated before the data were collected [...]." For illustration sake, imagine a researcher runs 20 regressions on the same dataset with income as response variable. The 20 regressions differ only in their explanatory variable. Now imagine no association exists between any of these 20 variables (body height, hair color, etc.) and income. Nonetheless, our "industrious" researcher has a 64.2%-chance of finding at least one statistically significant relationship among the 20 that were tested.[5] It is probably redundant to say at this point that what a significance-pursuing researcher might consider a chance to publish one more paper constitutes a serious damage to science because it leads to an inflation of false findings.

---

[5]Running a statistical significance test for 20 X-Y-relationships when they are all zero represents a Bernoulli experiment: one independently repeats 20 times a trial with two possible outcomes "significant" S and "not significant" $\overline{S}$, with $P(S) = 0.05 = p$, and $P(\overline{S}) = 0.95$. The probability of finding "significant" coefficients despite a fully random data structure is given by the binomial distribution $B_{m,p}(k)$, with $m = 20$ indicating the number of repeated significance tests, $p = 0.05$ the probability of falsely claiming a "significant discovery," and $k \in \{0, \ldots, 20\}$ the number of falsely claimed discoveries (Altman & Krzywinski, 2017; Hirschauer et al., 2018). The probability of finding no significant result despite the $m = 20$ repetitions is $B_{20,0.05}(0) = 35.8\%$. Expressed the other way round, the researcher has a "chance" of 64.2% of being able to proclaim at least one significant relationship even if all 20 relationships are non-existent.

6.2.5   Approaches Aimed at Mitigating Publication Bias

Many suggestions have been made to mitigate publication bias (see Munafò et al., 2017 for an overview). Song et al. (2013) and Weiß and Wagner (2011) propose to strengthen alternative publication outlets. They also call for a general change of editorial policies toward giving equal publishing chances to all scientific results, including replications and "negative findings." In addition, pre-analysis plans and the formal **preregistration** of complete data collection and study designs on a public repository *before* they are carried out are believed to reduce selective reporting (Olken, 2015; Munafò et al., 2017).

Pushing the issue of preregistration further, prominent economic societies started to **review pre-analysis plans** before publishing them. In 2013, the American Economic Association launched a preregistration scheme for randomized controlled trials (AEA, 2021b) in which a study's design is reviewed based on its methodological quality and officially registered if accepted. Reviewing a study's design and its formal registration by a prestigious institution are meant to prevent *p*-hacking and increase the chance of being published independent of whether positive or negative results are eventually found (Nosek et al., 2018). Preregistration, which is especially on the rise in the field of experimental economics, is increasingly recommended by other economics associations such as the European Economic Association.

Reaching still further and addressing the publishing decisions by reviewers and editors, some journals have adopted "**result-blind reviews**" (or, pre-result reviews; see Center for Open Science (2021) for an overview). A prominent example is the *Journal of Development Economics* that started testing in 2018 whether preregistration in conjunction with pre-result reviews can improve the quality of experimental economics research (Foster et al., 2019). In the pilot, the journal commits itself to publishing the results of studies whose designs had been successfully reviewed and preregistered by the journal.

Besides the various forms of preregistration, several other institutional attempts aim to counteract distorting selection procedures. In medical research, a global initiative *All Trials Registered/All Results Reported* was launched in 2013. Along those lines, the *Journal of Negative Results in BioMedicine,* the *PLOS Journals*, and the *All Results Journals* encourage replications and pursue **policies of publishing positive and negative results**. While data-sharing policies that facilitate replication in principle have been widely adopted in economics since their introduction by the AEA in 2005 (cf. AEA, 2021a), institutionalized efforts to strengthen replication and preregistration seem to be weak in economics compared to other fields such as the medical sciences. In a study of all 333 economic Web-of-Science journals, Duvendack et al. (2015) found that most of them still give very low priority to replication. Preregistration of studies also seems to lag behind other fields. As of 2017, no economic journal was among the approximately 40 journals that had adopted a policy of reviewing and registering study designs before results are known (Duvendack et al., 2017). But transparency is increasingly considered to be an important issue (see Miguel (2021) for an overview of transparency efforts in economics). There are also some **replication platforms** for economists such as *The Replication Network* and *Replication in Economics* that provide databases of replications and the opportunity to publish replication studies.

To avoid misunderstandings, one should finally note that testing multiple analytical variants and searching for potentially interesting patterns within a dataset is not per se a problem. On the contrary! But testing multiple analytical variants must be transparently reported as robustness check or labeled as **exploratory search**. Explorations often represent a primary step within the process of research that indicates what might be worth investigating (estimating) with new data in the future. We can understand p-hacking as a problem that arises when the explorative search for interesting patterns in a dataset is conceptually confused with the data-based estimation of a pre-defined population quantity of interest.[6] After the preliminary identification of, say, a potentially interesting association through an exploratory analysis, the question of whether the size of this association is relevant (substantially different from zero) in the parent population has to be pursued with a new set of data (Motulsky, 2014; Head et al., 2015). Otherwise, reasoning becomes a nonsense HARKing exercise that runs as follows: after an undirected explorative search in a dataset, researchers find a somehow interesting pattern—there will always be one if one searches long enough. Then, they "sell" the pattern in this particular dataset as evidence for its existence in the parent population. To avoid confusion, explorative research should always be transparently labeled as exploratory research.

6.3   The Historical Origins of the NHST-Framework

6.3.1   NHST: An Ill-bred Hybrid of Two Irreconcilable Statistical Approaches

Section 6.2 has shown that the NHST-procedure causes a substantial loss of information. It downgrades the two substantial pieces of information that we can derive from a random sample—the point estimate (signal) and the uncertainty of the estimation (noise) caused by random sampling error—first into a quotient (signal-to-noise ratio), and then into a p-value (based on the usually meaningless null hypothesis of zero effect), and finally into a dichotomous significance declaration (based on an arbitrary threshold). Nothing is gained by this downgrade. On the

---

[6]Section 6.2.2 has shown that the NHST terminology that speaks of "hypothesis *testing*" and "*confirmatory* analysis" is delusive because a p-value cannot be used to "test" or "confirm" a hypothesis in any usual sense of the word. It describes data frequencies under the null hypothesis, but it cannot work inversely and tell us whether a hypothesis is true or what is its probability. The very brevity of the established term "confirmatory analysis" has an advantage, however. It facilitates opposing "explorative analysis" to "confirmatory analysis" to indicate, in a very simple manner of speech, that the undirected search for potentially interesting patterns in a particular dataset must not be mixed up with the estimation of a pre-defined population quantity.

contrary. The dichotomization and the associated terminology that speaks of "significant" (or, "positive") as opposed to "non-significant" (or, "negative") results is dangerously deceptive. As a result, not only the consumers of research but also many researchers draw rash yes/no conclusions from findings of individual studies, instead of cautiously arguing what one should most reasonably believe in the light of the available evidence and the inevitably remaining uncertainty.

The poor track record of NHST as an aid to judgment gives rise to the question of why the relatively simple foundations of statistical inference—random sampling, random sampling error, and the sampling distribution of sample-based estimates—were distorted almost beyond recognition into statistical significance declarations. Unfortunately, the answer is neither simple nor definite. It seems that path dependence after a "wrong historical bend" has led to an amalgamation of two irreconcilable approaches that were originally designed for two very different contexts (Hirschauer et al., 2021a). An increasing desire for seeming unambiguousness backed up by ostensible mathematical exactitude seems to have contributed to the proliferation of the ill-bred hybrid "NHST" during the 20th century.

Looking back in history reveals that the NHST-framework is an amalgamation between the "**hypothesis testing**" approach by Neyman and Pearson (1933a, 1933b), on the one hand, and the "**significance testing**" approach by Fisher (1925), on the other (cf. Lehmann, 1993; Ziliak & McCloskey, 2008; Hurlbert & Lombardi, 2009; Kennedy-Shaffer, 2019). Inferential errors associated with NHST seem to be largely caused by a lacking familiarity with the original perspectives: Fisher focuses on inductive reasoning and on forming reasonable scientific beliefs ("inductive beliefs") from a given set of data. In contrast, statistical decision theory according to Neyman and Pearson aims at providing behavioral rules across repeated decisions ("inductive behaviors") under consideration of error costs.

Going back to Neyman and Pearson seems to be natural for economists in the tradition of decision theory, which deals with the probability distributions of costs and benefits caused by different actions. A decision-theoretic perspective clarifies that even when we succeed in identifying the most reasonable inductive belief given the available evidence, advisable courses of action cannot automatically be derived from such beliefs. As Berry (2017: 895) pointedly noted: "[. . .] believing or advertising something as true and acting as though it is true are very different kettles of fish." That is, we need to distinguish two different tasks: judging what we should most reasonably believe regarding a real-world state of interest as opposed to judging which decision we should most reasonably make after additionally considering the costs of wrong decisions. Obviously, we should shut down a nuclear power plant long before we have reason to believe that a meltdown is more likely than not.

Since the amalgamation of Fisher's approach and Neyman-Pearson's approach into the hybrid NHST ignores the crucial distinction between the above-mentioned "two kettles of fish," we scrutinize the two original approaches. Getting acquainted with these very different perspectives will clarify why cobbling them together virtually invites inferential misconceptions. Based on the understanding of the fundamental differences between Neyman-Pearson's and Fisher's perspective, we show that *moving forward* in statistical inference requires *going back to the roots* and using the standard error as measure of the uncertainty of sample-based estimates.

6.3.2 Inductive Behavior (Hypothesis Testing) and Type I Error Rates α

Statistical Decision Theory in the Neyman-Pearson Tradition
In statistical **hypothesis testing** (or, statistical decision theory) according to **Neyman and Pearson**, the world (technically speaking, the "parameter space") can presumably be divided into two mutually exclusive states—represented by the null and the alternative hypothesis (dichotomous perspective). In addition, constant conditions across many repeated decisions are assumed. This has several implications:[7]

1. One must not only specify a null hypothesis H_0 (implying that the alternative hypothesis remains a vague "non-null" proposition) but also a concrete alternative hypothesis H_A. Moreover, H_0 and H_A together are simplistically assumed to represent all possible states of the world.
2. A choice has to be made between these two hypotheses based on a statistical test. This test represents the "rule of behavior" for decisions that are made many times—each based on a new random sample obtained from an identical sampling design (e.g., in industrial quality control).
3. For each random sample, a "test score" (e.g., a z- or t-value) is computed based on the effect size and the standard error.
4. A test result that induces the decision associated with H_A (=rejection of H_0) or the decision associated with H_0 (=non-rejection of H_0) is *not* linked to an inductive belief that H_A or H_0 is true or more likely in the particular instance of testing. Instead, "accepting a hypothesis" means to act as if it were true in the light of the costs associated with either choice. This is also called "inductive behavior."
5. Different choices are fraught with different types of errors: type I errors arise if one acts as if H_A were true when in fact H_0 is true (false rejection of H_0). In contrast, type II errors arise if one acts as if H_0 were true when in fact H_A is true (false non-rejection of H_0).
6. The rule of behavior is based on an *a priori* fixed level of the type I error rate (false positive rate), which is usually designated by α. When the test results in a test score that leads to $p \leq \alpha$, one is to act as if H_A were true ("accept H_A"). When the test results in $p > \alpha$, one is act as if H_0 were true ("accept H_0").
7. Since the test is a "rule of behavior" aimed at guiding decisions that are made many times under constant conditions, the particular value of p in a particular test is *completely* irrelevant. The only relevant information is whether p falls into the rejection region or not.

---

[7]In regression analysis, α is often used for the intercept and β for the regression slope coefficient. We also used this notation in previous sections. In the current Sect. 6.3, we do not concern ourselves with regression analysis but focus on the estimation of a difference $\widehat{\delta}$ between the means of two groups. This enables us to keep with the conventional notation of statistical decision theory, which uses α for the type I error rate and β for the type II error rate.

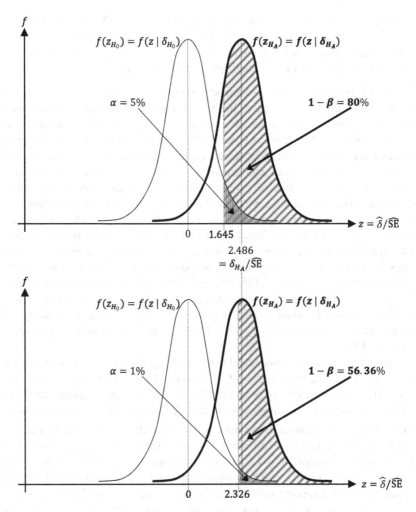

Fig. 6.5 The tradeoff between the type I error rate α and the type II error rate β

8. In subsequent samples and tests, p-values will be different. Therefore, a p-value found in a particular sample is not the type I error rate over many replications. But consistently following the rule of rejecting H_0 when $p \leq \alpha$ guarantees that, in the long run, the type I error rate will be α. This also follows from the fact that the p-value is uniformly distributed under the null hypothesis.

9. There is a tradeoff when setting the level of α: decreasing the type I error rate α ("false positive rate") across repeated decisions will necessarily increase the type II error rate β ("false negative rate") across those decisions (see Fig. 6.5). Expressed the other way round: lowering α decreases the power $1 - \beta$, i.e., the long-term rate of acting as if H_A were true when it is true ("true positive rate").

10. While using $\alpha = 0.05$ is often seen as a general default, Neyman and Pearson explicitly warned against a standard level for all decision contexts. Importantly, the magnitude of type I and type II error costs ("loss function") in a particular context must be considered when setting the level of α.

Figure 6.5 illustrates the relationship between the type I error rate α and the type II error rate β for a mean difference. In the upper part of the figure, we assume a decision rule $\alpha = 0.05$ and a sample size that is just large enough to make the standard error so small that we obtain $\beta = P(z < 1.645 \mid H_A) = 0.2$ for the given alternative hypothesis H_A. In other words, we have a power $1 - \beta = 0.8$. **Power** is the zeroth order *upper* partial moment of the z-distribution under the alternative hypothesis H_A for the value of z (also called "test statistic") associated with a particular α. This partial moment is sufficient in a dichotomous rejection/non-rejection context, i.e., power quantifies the repeatability of $p \leq \alpha$ when H_A is true.

The lower part of Fig. 6.5 illustrates the tradeoff between the type I error rate α and the type II error rate β. It describes what happens when we ceteris paribus lower the type I error rate to $\alpha = 0.01$, which corresponds to $z = 2.326$. This increases the type II error rate to $\beta = P(z < 2.326 \mid H_A) = 0.4364$, which corresponds to a power of $1 - \beta = 0.5636$. In contrast, if we accepted a higher type I error rate such as $\alpha = 0.1$ (not depicted), which corresponds to $z = 1.282$, we could lower the type II error rate to $\beta = P(z < 1.282 \mid H_A) = 0.1141$, which corresponds to a power of $1 - \beta = 0.8859$.

Figure 6.5 also helps us understand that increasing the sample size increases the signal-to-noise ratio z and shifts the z-distribution under H_A to the right. This lowers the type II error rate β for any predefined level of α. The term **"power analysis"** describes the search for a sample size n that, for a given H_A, a given α, and a presumably known variance in the data, produces a $\widehat{SE}$ that guarantees a desired power $1 - \beta$. Let's take a mean difference $\delta_{H_A} = 100$, $\alpha = 0.05$, and a desired power of $1 - \beta = 0.8$ as example: $z \geq 1.645$ is needed to meet $\alpha = 0.05$ under H_0. Power $1 - \beta = P(z \geq 1.645 \mid H_A) = 0.8$ is obtained when the z-distribution under H_A is centered around an expected value of 2.486, i.e., when $z_{H_A} \sim N(2.486; 1)$. This is achieved when $\widehat{SE} = 40.22$ ($2.486 = \delta_{H_A}/\widehat{SE} = 100/40.22$). Assuming that the standard deviation are $s_1 = s_2 = 744.5$ in both groups, $\widehat{SE} = 40.22$ is obtained according to Eq. (3.4) when the sample size is $n = 685$.

Statistical decision theory in the Neyman–Pearson tradition rejects the idea of forming a reasonable inductive belief about a real-world state of interest. Instead, statistical decision theory uses "hypothesis testing" to identify rules of behavior for two mutually exclusive decisions in the light of a **"loss function."** The dichotomous loss function by Neyman and Pearson considers the magnitude of type I and type II error costs. It also assumes zero costs both when we act as if H_0 were true when it is true ("true positive"), and when we act as if H_A were true when it is true ("true negative"). Neyman and Pearson (1933a: 296; 1933b: 497) emphasize that it "must be left to the investigator" to set an appropriate α that strikes the balance between the two types of errors "to meet the type of the problem before us."

Many vivid examples have been used to underpin the importance of considering type I and type II error costs when determining the decision rule α. An illustrative example is provided by Harvey (2017: 1408) who uses the comparison between "a jet engine failing" vs. "a water heating failing" to show how different the problems before us can be: "In the case of the jet engine, we are willing to accept a lot of false positives (incorrectly label a part defective) to minimize chances of false negatives (miss detecting a defective part), so α is set to a higher level. The particular situation therefore dictates not only how low α will be set but also the Type II error rate." Obviously, in the case of water heating, missing detecting a defective part is less fatal. Compared to the jet engine context, we would, therefore, accept more false negatives and less false positives.

The Performance of Statistical Decision Theory from a Decision-theoretic Perspective
Neyman and Pearson (1933a: 291) may have triggered off overly high expectations regarding the capacity of *statistical* **decision theory** to provide decision support by claiming: "Without hoping to know whether each separate hypothesis is true or false, we may search for rules to govern our behavior with regard to them, in following which we insure that, in the long run of experience, we shall not be too often wrong." Despite this claim and despite the use of the label "decision theory," the approach by Neyman and Pearson is *not* a decision-theoretic approach because it remains a conditional-probability concept: α is the long-term type I error rate *when H_0* is true $(P(type\ I\ error \mid H_0))$, and β is the long-term type II error rate *when H_A is true* $(P(type\ II\ error \mid H_A))$.

Since the Neyman-Pearson approach provides no scientific propositions regarding the probabilities of H_0 or H_A, a decisive piece of information is missing that is needed to obtain a rational normative rule from the point of view of **decision theory**. This holds even when we accept the dichotomous loss function by Neyman and Pearson and assume that type I and type II error costs are the only economically relevant factors that need to be considered in the decision. A decision rule that is to minimize expected costs needs to take into account *how often* we can expect to commit which type of error in the long run of testing. To obtain this information, we need not only to consider the **conditional probabilities** of type I and type II errors but also the **unconditional probabilities** of the two mutually exclusive states of the world described by the null and the alternative hypothesis.

It cannot be emphasized enough that a formal consideration of unconditional probabilities of states of the world is beyond the conditional probability concept of Neyman and Pearson. Expressed the other way round, the behavioral rule α in *statistical* decision theory will not bring about rational decisions unless we optimistically assumed that it were able to *implicitly* consider not only the relative magnitude of type I and type II error costs but also the probabilities $P(H_0)$ and $P(H_A)$ of the two mutually exclusive states of the world. Unfortunately, this optimism is not justified in view of the complex interactions of these factors. Even worse, the assumptions regarding the error costs and, even more so, regarding the state-of-the-world probabilities that should underlie the decision rule α are often not even

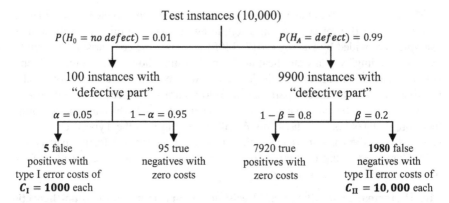

Fig. 6.6 Number of false decisions (false positives and false negatives) for a probability $P(H_0) = 0.01$ and a decision rule $\alpha = 0.05$ associated with $\beta = 0.2$

mentioned. Instead, $\alpha = 0.05$ and $\beta = 0.2$ are erroneously considered to be an adequate standard in all circumstances.

Figure 6.6 uses the "jet engine" example to illustrate the almost unaccomplishable task that we would face if we wanted to adequately "meet the type of the problem before us" based on the Neyman-Pearson approach alone. The example also shows that it would be pure coincidence if using $\alpha = 0.05$ and $\beta = 0.2$ as a default were correct. We assume that there are 10,000 instances where a risk-neutral decision-maker makes a decision by using the decision rule α. As in the upper part of Fig. 6.5, we assume that the sample size is just large enough to make the standard error so small that we obtain $\beta = 0.2$ when using the decision rule $\alpha = 0.05$.[8] As already shown in Fig. 6.5, this implies that the standard normal distribution of z under the null hypothesis ($z_{H_0} \sim N(0; 1)$) is accompanied by a normally distributed z under the alternative hypothesis ($z_{H_A} \sim N(2.486; 1)$).

We now presume type I error costs of $C_I = 1000 \,€$ and type II error costs of $C_{II} = 10,000 \,€$. Based on prior experience, we expect to see a defective part very often and correspondingly assume probabilities $P(H_0 = \text{no defect}) = 0.01$ and $P(H_A = \text{defect}) = 0.99$. In this situation, we would incorrectly label a part defective in only 5 of the 10,000 test instances. The total costs of the decisions associated with these five false alarms would amount to 5000 €. At the same time, we would miss detecting a defective part in 1980 of the 10,000 test instances. The costs of these misses would be 19,800,000 €.

In the situation described in Fig. 6.6, it is intuitively clear that we should use a decision rule α much larger than 0.05 in order to reduce β and thus the high costs of

---

[8]Taking the sample size as given implies that considering the antagonistic tradeoff between α and β, the magnitude of type I and type II error costs, and the probabilities $P(H_0)$ and $P(H_A)$ suffices to minimize long-term costs. Including sample size into the set of decision variables would additionally require considering that a costly increase of sample size n ceteris paribus decreases β. In this case, minimizing costs would imply finding the optimal combination of n and α.

missing detecting defective parts. To be more precise, we must identify the optimal decision rule α_{opt} that minimizes the expected value of costs $E(C_I) + E(C_{II})$ and thus the long-term costs over many repeated decisions. We use $F^z_{H_0}$ and $F^z_{H_A}$ to denote the cumulative distribution function of the z-distribution under the null hypothesis H_0 and the alternative hypothesis H_A. Using this notation, the optimization problem can be written as follows:

$$\min_{\alpha} \ [E(C_I) + E(C_{II})] = \alpha \cdot P(H_0) \cdot C_I + \beta \cdot P(H_A) \cdot C_{II},$$
$$\text{with } \beta = F^z_{H_A}\left(\arg\left(F^z_{H_0} = 1 - \alpha\right)\right) \qquad (6.1)$$

Equation (6.1) takes into account the antagonistic relationship between α and β, the type I and type II error costs, and the assumed probabilities of the two presumably exclusive states of the world. We obtain an optimal (cost-minimal) decision rule $\alpha_{opt} = 0.937$, through which the type II error is reduced to $\beta = 0.00003$. From a decision-theoretic point of view, *this* is an adequate consideration of "the problem before us." It reduces the total costs over 10,000 decisions to 96,623 €—compared to costs of 19,805,000 € that would arise with a "default" $\alpha = 0.05$ and $\beta = 0.2$.

The upper left field of Table 6.1 takes up the constellation described above. Contrasting it with the lower left field of Table 6.1 shows that things are very different if, based on prior evidence, it is reasonable to assume $P(H_0) = 0.99$, i.e., if we expect to encounter the state of the world of "no defective part" in 99% of all test instances. In this case, we can expect to make 495 type I errors and only 20 type II errors over the 10,000 tests when we use the decision rule $\alpha = 0.05$ accompanied by $\beta = 0.2$. The costs of false alarms (expected type I error costs) would total 495,000 € and the costs of missing a defective part (expected type II error costs) would only total 200,000 €. Given the reversed probabilities, it is intuitively clear that we should

Table 6.1 Optimal decision rule α_{opt} and total cost over 10,000 test instances depending on various state-of-the-world probabilities and type I and II error costs[a,b]

	Type I error costs C_I: 1000 € Type II error costs C_{II}: 10,000 €	Type I error costs C_I: 10,000 € Type II error costs C_{II}: 1000 €
$P(H_0) = 0.01$ $P(H_A) = 0.99$	Default total costs: 19,805,000 € Minimal total costs: 96,623 € $\alpha_{opt} = 0.937; \beta = 0.00003$	Default total costs: 2,030,000 € Minimal total costs: 524,344 € $\alpha_{opt} = 0.374; \beta = 0.015$
$P(H_0) = 0.50$ $P(H_A) = 0.50$	Default total costs: 10,250,000 € Minimal total costs: 2,629,279 € $\alpha_{opt} = 0.376; \beta = 0.015$	Default total costs: 10,250,000 € Minimal total costs: 2,629,279 € $\alpha_{opt} = 0.015; \beta = 0.376$
$P(H_0) = 0.99$ $P(H_A) = 0.01$	Default total costs: 695,000 € Minimal total costs: 524,355 € $\alpha_{opt} = 0.015; \beta = 0.376$	Default total costs: 4,970,000 € Minimal total costs: 97,313 € $\alpha_{opt} = 0.000009; \beta = 0.964$

[a]Assumptions and notation: 10,000 test instances; sample size guarantees that default $\alpha = 0.05$ is accompanied by $\beta = 0.2$; H_0=no defective part; H_A=defective part
[b]Computation of costs as follows: Costs $= (\alpha \cdot P(H_0) \cdot C_I + \beta \cdot P(H_A) \cdot C_{II}) \cdot 10,000$. For example: $19,805,000 = (0.05 \cdot 0.01 \cdot 1000 + 0.2 \cdot 0.99 \cdot 10,000) \cdot 10,000$

lower α and accept an increase of β from a long-term cost perspective. The optimal decision rule is $\alpha_{opt} = 0.015$. This corresponds to $\beta = 0.376$ and leads to minimal total costs of 524,355 €, compared to costs of 695,000 € that would be incurred by the default decision rule $\alpha = 0.05$.

For Laplace probabilities $P(H_0) = P(H_A) = 0.5$ (see middle left field of Table 6.1), we obtain $\alpha_{opt} = 0.376$, which is in between the two extremes in the first and the third row. Rearranging Eq. (6.1) enables us to identify constellations where $\alpha = 0.05$ would be optimal. For example, Laplace probabilities in conjunction with a cost ratio $C_I/C_{II} = 2.714$ is a "coincidental" constellation where indeed $\alpha = 0.05$ (accompanied by $\beta = 0.2$) would be optimal. This emphasizes that using $\alpha = 0.05$ and $\beta = 0.2$ as default for *all* circumstances does not represent a rational rule of behavior. This holds even if we assume Laplace probabilities for the two mutually exclusive states of the world due to lacking prior knowledge.[9]

Let us now take a brief look at the right column of Table 6.1. Whereas we assumed type I error costs $C_I = 1000$ and type II error costs $C_{II} = 10,000$ in the left column, we now assume $C_I = 10,000$ and $C_{II} = 1000$. We intuitively guess that this reversion of the cost ratio makes us accept much less false alarms. In fact, we see that it causes a substantial decrease of the optimal type I error rate α_{opt} in all three constellations of state-of-the-world probabilities.

It is particularly instructive to compare the upper left field of Table 6.1 with the lower right field. The upper left field describes a situation where we have not only a high probability of seeing a defective part ($P(H_A = \text{defect}) = 0.99$) but also high costs of missing detecting a defective part. Therefore, we accept a lot of type I errors and set the decision rule to a very high level $\alpha_{opt} = 0.937$ to reduce the type II error rate to a very low level $\beta = 0.00003$. The lower right field describes exactly the opposite. Here we have not only a low probability of seeing a defective part ($P(H_A = \text{defect}) = 0.01$) in the first place but also low costs of missing detecting a defective part. We can thus reduce α to a very low level of $\alpha_{opt} = 0.000009$ and accept a high type II error rate $\beta = 0.964$.

The jet engine example above illustrates the shortcomings of the Neyman-Pearson approach. On the one hand, a rational decision rule cannot be based on conditional probabilities alone. To support rational choices, one needs a decision-theoretic framework that additionally considers the unconditional state-of-the-world probabilities and, where applicable, the individual risk attitude. On the other hand, the Neyman–Pearson approach also does not help answer the question of what we can reasonably claim to learn regarding a real-world state of interest from a given set of data. This is because, being *meant* to yield a decision rule, it accounts for choice-related costs. While rational decisions depend on the costs that are associated with different courses of action, reasonable beliefs do not.

---

[9]Distributional and specification tests (e.g., Shapiro–Wilk test, Hausman test) are also tests whose results induce choices—methodological ones, now. While a decision rule $\alpha = 0.05$ is routinely used, α should be specified according to the above rationale, i.e., under consideration of prior probabilities and the type I and II error costs associated with wrong methodological choices.

6.3.3 Inductive Belief (Significance Testing) and p-Value Thresholds

Section 6.3.2 has clarified that the Neyman–Pearson approach, which attempts to provide decision support under consideration of error costs, does not provide any help for forming inductive beliefs. What remains to be done to understand the inconsistency of the hybrid NHST is to briefly look into the original **significance testing** approach by **Fisher**.

Even though NHST has borrowed the term "hypothesis testing" from statistical decision theory, it is closer in spirit to Fisher's view that the p-value represents a helpful tool in the difficult exercise of forming inductive beliefs. Most econometricians would probably agree with Fisher's (1935: 39) statement of what this exercise is about: "[...] everyone who does habitually attempt the difficult task of making sense of figures is, in fact, essaying a logical process of the kind we call inductive, in that he is attempting to draw inferences from the particular to the general." In other words, NHST has not adopted the Neyman-Pearson perspective of providing a behavioral rule under consideration of error costs. Instead, it shares Fisher's interest of forming inductive beliefs from a given set of data. This also coincides with our interest to assess what we can reasonably claim to have learned from a data-derived signal (point estimate) and its noise (standard error).

The practical approach of NHST in econometrics is also more similar to Fisher's than to Neyman–Pearson's perspective in that the world is not dichotomized into two mutually exclusive states represented by a null and one alternative hypothesis. But being a misleading hybrid, many NHST-based studies come up with inductive yes/no propositions depending on whether $p \leq 0.05$ or not. While Fisher emphasized that inductive inferences always remain uncertain, he unfortunately did suggest labeling results associated with a p-value less than or equal a threshold of 0.05 as "statistically significant." He even claimed that "non-significant" results can be ignored. This seems to have contributed to the dissemination of dichotomous language and thinking in inductive inference and the ensuing rise of NHST.

Fisher's clarification that low p-values just *signify* "worth a second look" and later warnings that, beyond "convenient" significance statements based on a threshold such as 0.05, "exact p-values" indicating the strength of evidence against the null hypothesis should be used as an aid to judgment (Fisher, 1960: 25) went largely unheeded. In any case, it didn't prevent the rise of the hybrid NHST, which is neither in line with Neyman–Pearson's nor Fisher's approach. A particularly devastating confusion stems from the fact that NHST does not distinguish between Neyman–Person's *behavioral rule* α (often also set to a default of 0.05) and Fisher's *convenient threshold* of $p = 0.05$. This is inconsistent since a test in Neyman–Pearson's approach is not used to obtain an inductive belief regarding a hypothesis and since the exact value of p is completely irrelevant. The opposite is true for Fisher's approach which, according to his later warning, does not need the label "significance" but should use the exact p-value as statistical aid for forming reasonable beliefs.

Even if we followed Fisher and used exact *p*-values to support our inferences, there is only one thing that can be said: small *p*-values represent stronger evidence against the point null hypothesis than large *p*-values because small *p*-values occur more often if there is an effect compared to no effect (cf. Sect. 6.1). Another way of expressing the meaning of a *p*-value would be to say that lower *p*-values indicate a higher incompatibility of the data with the point null hypothesis. But as already mentioned, even exact *p*-values represent an information downgrade compared to the two original pieces of information that we can derive from a random sample, i.e., the point estimate (signal) and the standard error, which quantifies the uncertainty of the estimation (noise) caused by random sampling error. This downgrade—first into a quotient (signal-to-noise ratio) and then into a *p*-value (based on the usually meaningless null hypothesis of zero effect)—brings about a substantial loss of information. Therefore, reporting *p*-values provides a less suitable aid for making adequate inferences than separately reporting signal and noise. This holds even if we could assume that *p*-values were properly interpreted.

Further downgrading *p*-values into dichotomous significance declarations produces an even more severe loss of information. What is worse, the NHST-routine with its hypothesis testing terminology and dichotomous significance declarations propagate cognitive biases and virtually seduce researchers to make logically inconsistent and overconfident inferences, both when *p* is below and when it is above the "significance" threshold. These misinterpretations are rooted in the very fabric of NHST as an inconsistent hybrid of the "hypothesis testing" approach by Neyman and Pearson and the "significance testing" approach by Fisher. The amalgamation is inconsistent because it pretends that we can use a signal-to-noise ratio (e.g., 1.645) to obtain a dichotomous rule of what to believe, while such a dichotomy would, at best, make sense as a rule of what to do when facing either-or choices with presumably known error costs. As a consequence, it invites the misunderstanding that inferences flow from data automatically.

Unfortunately, statistical practitioners have widely succumbed to NHST's misleading invitation and forgotten that the standard error, upon which all inferential statistics are based, is no more than an estimate of the standard deviation of the sampling distribution. Because of dichotomous significance declarations, researchers all too often ignore that what they have is a signal and a noise information. Consequently, common interpretations in the NHST-framework deviate substantially from Fisher's modest "worth a second look" interpretation of low *p*-values. While we can reduce uncertainty through larger samples and assess uncertainty by correctly estimating the standard error, we must embrace the random-error-related uncertainty as long as we study samples instead of full populations.

In brief, empirical research in the social sciences is usually not directly aimed at providing decision support. Rather, its objective is to gain incremental knowledge that helps us form the most reasonable **inductive belief** (proposition) regarding a real-world state of interest in the light of the available evidence. Forming plausible beliefs is a precondition for making rational decisions but it is an independent working step. As such, it must be clearly separated from decision-oriented approaches such as the Neyman–Pearson approach, which *attempts* to arrive at an

inductive behavior (a behavioral rule) given the costs of wrong decisions. Dichotomies and thresholds such as the decision rule α used in "hypothesis testing" are not only superfluous but seriously misleading in contexts where we weigh the evidence to answer the question of what we can reasonably claim to learn from a given set of data. What we should most reasonably believe does not depend on the costs that are related to different courses of action, but rational decisions certainly do. To be more precise, without presuming that a choice has to be made between two mutually exclusive courses of action under consideration of the expected type I and type II error costs, dichotomous hypothesis testing is meaningless from the very start.

6.4   Chapter Summary: Significance Declarations Should Be Avoided

We have not been digging very deep into the history of statistical science, but we hope that this primer on the differing perspectives of inductive reasoning (Fisher) as opposed to statistical decision theory (Neyman and Pearson) helps the reader prevent inferential errors that are largely due to the NHST-amalgamation of these irreconcilable approaches. The low performance of NHST as an auxiliary tool for making inductive inferences from the sample toward a broader population brings us back to the basics of statistical inference as described in Chaps. 3, 4, and 5: the original information that can be derived from a random sample is, at best, an unbiased point estimate (signal) and an unbiased estimation of the uncertainty of that point estimation (noise) expressed through the standard error (i.e., the estimated standard deviation of the sampling distribution). These two quantities are intelligible and meaningful if there was a probabilistic data generation process. Downgrading them first into a quotient (signal-to-noise ratio), and then into a p-value (based on the usually meaningless null hypothesis of zero effect), and finally into a dichotomous significance declaration (based on an arbitrary threshold) is not by itself wrong. We can, of course, perform the underlying mathematical manipulations. The procedure is not useful, however. It causes not only a substantial loss of information but virtually provokes overconfident conclusions, both when p is below and when it is above the "significance" threshold.

Very early in this book, the drawbacks of NHST were already anticipated by the gender pay gap example in Table 3.1. Recall that the two studies obtained the same mean pay gap estimate $\widehat{\delta} = 100$. But due to a ten times larger sample size, the estimation uncertainty caused by random sampling error was much smaller in study 1 ($\widehat{SE} = 60.79$) compared to study 2 ($\widehat{SE} = 192.23$). Nothing is gained—except seemingly neat but nonetheless wrong yes/no conclusions—when the sample-derived information (i.e., the pay gap estimate and its standard error) is downgraded to the statement that a "statistically significant" pay gap ($p = 0.050$) was observed in study 1 and a "statistically non-significant" pay gap ($p = 0.303$) in study 2.

Given the poor track record of *p*-values and NHST as an aid to judgment, it only remains recalling that *moving forward* in statistical inference requires *going back to the roots* and using the standard error as measure of the uncertainty of an estimation. For the sake of transparency and effective scientific communication, empirical researchers should therefore keep with the following procedure of statistical inference, which we are in principal already familiar with from Sect. 3.4:

1. Unambiguously define the parent population from which the random sample was drawn. Be clear that the parent population defines the inferential target population to which inductive *statistical* inference is limited.
2. Define the probabilistic sampling design (e.g., SRS, stratified sampling, cluster sampling) that was used to collect the sample from the defined parent population.
3. First describe the empirical evidence you found in your specific dataset (e.g., a sample mean, a sample mean difference, or a sample regression slope) and then transparently communicate that you want to use the observed sample quantity as a point estimate (signal) for the population quantity of interest.
4. Estimate the standard error, i.e., the uncertainty of the point estimation caused by random sampling error (noise). Be clear that the standard error formula depends not only on the type of the estimate (mean, mean difference, regression slope) but also the sampling design (SRS, stratified sampling, cluster sampling).
5. Finally, tackle inductive inference and assess the validity of the observed sample quantity as a point estimate (signal) for the population quantity of interest. Use the two intelligible pieces of information derived from the sample—the point estimate (signal) and its uncertainty (noise)—in a comprehensive scientific reasoning that makes reasonable inferences from the idiosyncratic random sample toward the real-world state of interest in the light of all available information.

Do not overrate the informative value of statistical inference: first, inductive inferences never flow from data automatically and inferential statistics are just *auxiliary* tools for making reasonable judgments. Second, the validity of findings beyond the confines of an idiosyncratic dataset is impaired by more than just *random* sampling error. Other types of error often cause more problems for the "inferential leap" from a particular set of data to the social group and real-life setting of interest. Selection bias (*non-random* sampling error) that cannot be corrected for is an important example. Therefore, be clear that estimation uncertainty caused by random error is the only type of uncertainty the standard error and derived statistical quantities deal with. Third, *statistical* inference based on the signal and the noise from a single study is only one part of the much larger enterprise of scientific inference. *Scientific* inference means drawing reasonable conclusions regarding a real-world state of interest given all prior knowledge (e.g., from previous studies) *and* the incremental information that was extracted from the particular sample. In this larger enterprise of accumulating evidence and knowledge, statistical inference has the role of helping evaluate the single study's knowledge contribution. In this evaluation, the standard error is an important means to assess how much knowledge gain is achieved given the "noise" from random sampling error.

6.5   Recommended Reading

Amrhein, V., Greenland, S., & McShane, B. (2019). Retire statistical significance. *Nature, 567*, 305–307.

Brodeur, A., Lé, M., Sangnier, M., & Zylberberg, Y. (2016). Star wars: The empirics strike back. *American Economic Journal: Applied Economics, 8*(1), 1–32.

Cohen, J. (1994). The earth is round (p < 0.05). *American Psychologist, 49*(12), 997–1003.

Denton, F. T. (1988). The significance of significance: Rhetorical aspects of statistical hypothesis testing in economics. In A. Klamer, D. N. McCloskey, & R. M. Solow (Eds.), *The consequences of economic rhetoric* (pp. 163–193). Cambridge University Press.

Gigerenzer, G., & Marewski, J. N. (2015). Surrogate science: The idol of a universal method for statistical inference. *Journal of Management, 41*(2), 421–440.

Greenland, S., Senn, S. J., Rothman, K. J., Carlin, J. B., Poole, C., Goodman, S. N., & Altman, D. G. (2016). Statistical tests, P values, confidence intervals, and power: A guide to misinterpretations. *European Journal of Epidemiology, 31*(4), 337–350.

Hirschauer, N., Grüner, S., Mußhoff, O., & Becker, C. (2018). Pitfalls of significance testing and *p*-value variability: An econometrics perspective. *Statistics Surveys, 12*, 136–172.

Hirschauer, N., Grüner, S., Mußhoff, O., & Becker, C. (2019). Twenty steps towards an adequate inferential interpretation of *p*-values in econometrics. *Journal of Economics and Statistics, 239*(4), 703–721.

Hirschauer, N., Grüner, S., Mußhoff, O., & Becker, C. (2021a). A primer on p-value thresholds and α-levels – Two different kettles of fish. *German Journal of Agricultural Economics, 70*, 123–133.

Imbens, G. W. (2021). Statistical Significance, *p*-values, and the reporting of uncertainty. *Journal of Economic Perspectives, 35*(3), 157–174.

Krämer, W. (2011). The cult of statistical significance – What economists should and should not do to make their data talk. *Schmollers Jahrbuch, 131*(3), 455–468.

Lehman, E. L. (1993). The Fisher, Neyman-Pearson theories of testing hypotheses: One theory or two? *Journal of the American Statistical Association, 88*, 1242–1249.

McCloskey, D. N., & Ziliak, S. T. (1996). The standard error of regressions. *Journal of Economic Literature, 34*(1), 97–114.

National Academies of Sciences, Engineering, and Medicine. (2019). *Reproducibility and replicability in science. Consensus study report.* The National Academies Press. https://doi.org/10.17226/25303

Nickerson, R. S. (2000). Null hypothesis significance testing: A review of an old and continuing controversy. *Psychological Methods, 5*(2), 241–301.

Wasserstein, R. L., & Lazar, N. A. (2016). The ASA's statement on *p*-values: Context, process, and purpose. *The American Statistician, 70*(2), 129–133.

Wasserstein, R. L., Schirm, A. L., & Lazar, N. A. (2019). Editorial: Moving to a world beyond "$p < 0.05$". *The American Statistician, 73*(Suppl 1), 1–19. [The full material of the special issue "Statistical Inference in the 21st Century: A World Beyond $p < 0.05$" can be accessed at: https://www.tandfonline.com/toc/utas20/73/sup1]

Ziliak, S. T., & McCloskey, D. N. (2008). *The cult of statistical significance. How the standard error costs us jobs, justice, and lives.* The University of Michigan Press.

Chapter 7
Statistical Inference in Experiments

7.1 Inferential Cases in Group Mean Comparisons

The previous chapters focused on the role of statistical inference in observational studies based on **random sampling**. That is, we dealt with **generalizing statistical inference** (sample-to-population inference) and described how statistics can help us evaluate what a quantity such as a mean or a mean difference observed in a random sample can tell us about the population. We could also say that we were concerned with the **external validity** of a population estimate obtained from *one* random sample in the light of the estimation uncertainty caused by **random sampling error**. This uncertainty is estimated by the standard error, which is but another label for the standard deviation of the **(sampling) distribution** of the estimates that we would obtain over frequently repeated random draws from the population.

The present chapter focuses on statistical inference in experimental studies based on **randomization**. Such "randomized controlled trials" (RCTs) resort to a deliberate intervention (treatment) and a random assignment of experimental subjects to treatments. That is, we now focus on **causal statistical inference** and show how statistics can help us evaluate what a mean difference observed in a random assignment can tell us about the causal treatment effect. We could also say that we are concerned with the **internal validity** of a causal effect estimate obtained from *one* random assignment in the light of the estimation uncertainty caused by **randomization error**. Estimation uncertainty is, again, expressed through the standard error, but it now describes the estimated standard deviation of the **(randomization) distribution** of the causal effect estimates that we would obtain over frequently repeated random assignments of the given experimental subjects to treatments.

Table 7.1 distinguishes the inferential cases that arise in mean comparisons from the two dimensions of data generation: the selection mechanism for recruiting subjects and the origin of the group membership. **Case (1)** describes a situation where neither generalizing nor causal statistical inference is feasible because there is neither random sampling nor randomization. **Case (2)** refers to (non-randomized)

N. Hirschauer et al., *Fundamentals of Statistical Inference*, SpringerBriefs in Applied Statistics and Econometrics, https://doi.org/10.1007/978-3-030-99091-6_7

Table 7.1 Inferential cases in mean comparisons depending on data generation

	Non-randomization: subjects' membership in pre-existing groups (e.g., male/female) is observed	**Randomization:** subjects are randomly assigned to experimental treatment groups (RCTs)
Non-random selection of study subjects (convenience sample)	(1) *Neither generalizing inference nor causal inference* can be supported by *statistics*	(3) *Causal inference* regarding a treatment effect can be supported by *statistics*
Random selection of study subjects (random sample)	(2) *Generalizing inference* from the sample to the population can be supported by *statistics*	(4) *Causal inference and generalizing inference* can be supported by *statistics*

observational studies based on random sampling. Here, generalizing sample-to-population inference can be supported by statistics, but no causal statistical inference is feasible. **Case (3)** describes RCTs in which non-randomly recruited subjects are randomly assigned to treatments. Here, causal inference can be supported by statistics, but no generalizing statistical inference beyond the pool (sample) of experimental subjects is feasible. **Case (4)** refers to research designs where we first randomly recruit subjects from a population and then randomly assign them to different experimental treatments. From a statistical perspective, this is an ideal setting that facilitates generalizing causal statistical inference.

While being questionable on statistical grounds, p-values for mean differences are commonly displayed even when there is neither random sampling nor randomization, as in case (1). This raises the question of which kind of inference p-values are to support in such circumstances. Imagine an observational study that investigates gender-related pay patterns based on the payroll data of a company's male and female employees. Generalizing *statistical* inference is not possible because the employees were not randomly drawn from a larger population (cf. Fig. 2.1). Causal statistical inference is not feasible either because group allocation was not brought about by randomization. So what could a p-value possibly convey?

It cannot be stressed enough that one needs to *imagine* a chance mechanism for inferential statistical procedures to make sense when there is neither random sampling nor randomization. For example, one could imagine *fictitious random sampling* from a *fictitious superpopulation* (cf. Sect. 4.5). Since this is often not meaningful, researchers should spell out which kind of argument they want to support when resorting to an imaginary random process. Unfortunately, the logical necessity of a chance mechanism is often ignored and conventional inferential content is delusively suggested even when there is no chance mechanism. This holds especially in group comparisons where standard errors and p-values are often more or less automatically displayed. We therefore included a description of this apparently confusing issue (cf. Sect. 7.3) in the present chapter on experiments.

7.2   Causal Inference

7.2.1   Overview of Experimental Designs Aimed at Establishing Causality

The causal inferences that can be made from controlled experiments where everything but the treatment is held constant are different from those that can be made from observational studies. The former rely on the research design to ex ante ensure **ceteris paribus conditions** that facilitate the estimation of causal effects. This is known as "ex ante control of confounders." Observational studies aimed at establishing causal effects, in contrast, rely on an ex post control of confounders (e.g., through regression modeling) that does not provide a way of ascertaining causal relationships as reliable as an ex ante research design (Athey & Imbens, 2017).

Unfortunately, there is no generally acknowledged, inference-related systematization of experiments (Hirschauer et al., 2020a). Some scholars use randomization as the defining quality and equate "experiments" with RCTs (Athey & Imbens, 2017). Others include non-randomized designs into the definition as long as there is a deliberate treatment manipulation (Harrison & List, 2004; cf. Excursus). Confusion regarding the different research designs that are labeled as experiments entails the risk of inferential errors. Regarding the potential of statistical inference in experiments, two different designs need to be distinguished:

- In RCTs with their **randomized-treatment-group comparisons** (between-subject designs), causal inference can be supported by probability theory and statistics because control of confounders is obtained by randomly assigning subjects to treatment groups. Usually, the mean outcome of a control group (non-treatment group) is compared to the mean of a treatment group. Randomization generates equivalence between groups by ensuring that confounders are balanced across groups.[1] This is also known as "statistical independence of treatments."
- In **before-and-after-treatment** comparisons (within-subject designs), causal inference has nothing to do with probability theory and statistics. This is because all subjects are exposed to the treatment, and the individual outcomes before-and-after the treatment are then observed and compared. Rather than relying on a probabilistic procedure that balances confounders across treatment groups, control of confounders must be obtained by physically holding everything but the treatment constant over time.

---

[1] Randomization only achieves that confounders are balanced across treatments "in expectation" (Duflo et al., 2007), i.e., balance is only ensured in "sufficiently" large groups. In small groups, a difference between control and treatment could be (partly) due to unbalanced confounders. This, precisely, is taken account of by the statistical procedures that deal with the randomization-related uncertainty (Ramsey & Schafer, 2013: Chap. 1).

The persuasiveness of causal claims depends on the credibility of the alleged control of confounders. Randomization is generally held to be a more convincing device to identify causal effects than before-and-after treatment comparisons (Charness et al., 2012) because it balances known *and* unknown confounders across groups. That is, the treatment effect is a "between-subject variable" because we can only observe a subject's behavior under the treatment or in the absence of the treatment, but not both. This "missing data problem" (cf. Rubin, 2010) is solved by using the observed average outcome of the treatment group as counterfactual for the unobserved average outcome of the control group would and vice versa. In contrast, control of confounders in before-and-after-treatment comparisons is limited by the researcher's ability to identify and fix all relevant confounders. In other words, the outcome is a "within-subject variable" and causality claims hinge on the credibility of the ceteris paribus condition as substantiated by the research protocols that describe what was done to hold everything but the treatment constant.

Within-subject designs also have advantages (Czibor et al., 2019): first, they make more effective use of any given number of experimental subjects in that they generate treatment observations for all of them. Second, rather than being limited to comparing the means (or higher moments) of the outcome distribution in the control and the treatment group (for an overview of distributional impact analysis in RCTs, see Bedoya et al., 2017), within-subject designs enable researchers to observe individual treatment effects. This is an important advantage (Gelman, 2019). Imagine a treatment that causes benefits for some individuals but harms others. In an RCT, we would only see "net effects." In extreme cases, the outcome distribution under control and treatment might be identical despite numerous and substantial treatment effects on the individual level. In contrast, within-subject designs would reveal the (distribution of) individual harms and benefits from the treatment.

Insofar as they succeed in ensuring effective ex ante control of confounders, both RCTs and experiments based on before-and-after-treatment comparisons facilitate causal inferences. However, the role of statistical inference and the meaning of the standard error are completely different in the two cases. In RCTs, the standard error deals with randomization error and describes the **standard deviation of the randomization distribution**, i.e., the distribution of the differences in mean outcomes between control and treatment that we would find if we reran the experiment in many other randomizations (re-groupings) of the same pool of experimental subjects. In contrast, using the standard error in experiments based on before-and-after-treatment comparisons presupposes that the experimental subjects are a random sample of a broader parent population.[2] In this case, the standard error describes the **standard deviation of the sampling distribution** and quantifies the dispersion of the mean individual before-and-after differences that we would observe with the given

---

[2] We should recall at this point that we rarely have random samples because people can usually freely decide whether to participate in a study or not. Even when we are in the comfortable position of having access to the full population or a fully-random sampling frame, participants and non-participants are likely to be systematically different due to selection bias. The observations in the sample may thus tell us little about the population (cf. Sect. 4.4).

experimental design in many other random samples drawn from the same population. That is, it deals with random sampling error and generalizing inference but has nothing to do with the question of whether the experimental design is capable of holding everything but the treatment constant over time. In brief, the sampling-based standard error in experimental before-and-after comparisons contains no information regarding the uncertainty of the *causal* effect estimation.

In the remainder of this chapter, we will not further discuss statistical inference in experiments based on before-and-after-treatment comparisons because random sampling and generalization toward populations have already been dealt with in the previous chapters. In fact, there is nothing new in this regard except that the sample observations are now the observations of the variable "individual before-and-after difference in the experiment" instead of a pre-existing individual characteristic. We can consequently resort to Eqs. (3.1) and (3.2) to estimate the mean of this variable and its corresponding standard error.

Excursus: Besides within-subject and between-subject designs, which are aimed at identifying causal treatment effects, the term "experiment" is also used when researchers use an intervention but are not interested in establishing a causal effect. Such "experiments" are basically measurement devices to obtain (manifest) behavioral observations that are to reveal subjects' (latent) preferences. Experimental designs that credibly ensure that subjects' choices in the more or less artificial experimental setting reveal their "true" preferences are sometimes said to "exercise control over subjects' preferences" (Smith, 1982). Generally, economists consider incentives, which ensure that the choices in the experiment have substantial real consequences, as a prerequisite for an effective control over subjects' preferences.

Examples of "experiments" serving as measurement devices are multiple price lists for eliciting risk preferences or experimental games aimed at identifying social preferences. The quality of the control over subjects' preferences (and thus the quality of the measurement of the latent preferences) is an aspect of external validity (ecological validity) that has nothing to do with statistical inference. While we may jointly have randomization *and* random sampling *and* control over subjects' preferences, we may also have an "experiment" without random assignment and without random recruitment, but with a credible control over subjects' preferences. In this case, neither causal inferences nor generalizing inferences can be supported by statistics.

Imagine a well-incentivized dictator game aimed at identifying altruistic preferences is carried out with a convenience sample of students who happen to be in a researcher's classroom on a particular Friday. Random assignment to different treatments being absent, causal inferences are neither intended nor possible; and all generalizing inferences—whether they are toward the experimental behavior of a broader population of students or other demographic

(continued)

groups, or toward the real-life behavior of the classroom students or a broader population—must be based on scientific arguments beyond statistics. It would be a gross abuse to report p-values and "statistical significance" for a purported corroboration of these inferences in such a context.

7.2.2 The Uncertainty of Causal Effect Estimates Caused by Randomization

Let us take a closer look at RCTs and causal statistical inference. While we generally assume "reasonably large" samples and groups ($n \geq 30$) throughout this book, we explain the concept of randomization-based uncertainty by using a highly stylized example with a very small pool of $n = 4$ experimental subjects. Imagine these four subjects are randomly assigned to two equal-sized groups of two. The members of both groups are asked to act as first "players" in a dictator game and decide which share of an initial endowment of eight monetary units they give to an unknown second "player." The control group is simply asked to make the decision. The treatment group is shown a movie emphasizing that being altruistic is an endearing character trait before being asked to decide how much to pass on to the second player.

Let us assume that there is zero treatment effect ($\delta = 0$), but that the $n = 4$ subjects are naturally heterogeneous regarding the outcome of interest, i.e., the amount x they pass on to the second player ($x_1 = 1$, $x_2 = 2$, $x_3 = 3$, $x_4 = 4$). There are $\binom{n}{k} = \frac{n!}{k! \cdot (n-k)!} = \frac{4!}{2! \cdot 2!} = 6$ possible ways (**permutations**) to choose $k = 2$ subjects out of a group of $n = 4$. Consequently, there are six ways of randomly assigning the four subjects to two equal-sized groups (control and treatment).

Table 7.2 describes the six possible randomizations (permutations). Due to the vagaries of randomization, more or less subjects with naturally higher outcomes may be assigned to the one or the other group. Across all possible re-groupings, we would observe mean differences between treatment and control group that range from $d = 2$ to $d = -2$. Since we assumed the treatment effect to be zero ($\delta = 0$), the dispersion of the average treatment effect estimates $d = \widehat{\delta}$ that we would obtain in the different random assignments (i.e., the dispersion of the "exact" randomization distribution) is solely caused by randomization error $e_{\text{ran}} = \widehat{\delta} - \delta = \widehat{\delta}$.

While we refrain from calculating a standard deviation (standard error) for this discrete randomization distribution with six permutations only, the last two columns of Table 7.2 describe its dispersion by displaying quantiles. For example, due to the vagaries of randomization, the probability of obtaining an estimate that exceeds the true effect size by 1 or more amounts to 0.333 if there is no treatment effect.

Table 7.2 Possible randomizations for two equal-sized-groups with $n = 4$ experimental subjects and a presumably known treatment effect of $\delta = 0$

		Obser-vations	Means	Mean differences $d = \hat{\delta}$	Randomization error $e_{\mathrm{ran}} = \hat{\delta} - \delta$	(Reversed) cumulative probability
1	Control	1; 2	1.5	2.0	2.0	0.167
	Treatment	3; 4	3.5			
2	Control	1; 3	2.0	1.0	1.0	0.333
	Treatment	2; 4	3.0			
3	Control	1; 4	2.5	0.0	0.0	0.500
	Treatment	2; 3	2.5			
4	Control	2; 3	2.5	0.0	0.0	0.667
	Treatment	1; 4	2.5			
5	Control	2; 4	3.0	−1.0	−1.0	0.833
	Treatment	1; 3	2.0			
6	Control	3; 4	3.5	−2.0	−2.0	1.000
	Treatment	1; 2	1.5			

Enumerating all possible permutations for a small group provides an intuitive illustration of how randomization error makes causal effect estimation uncertain. However, in practical research, the usefulness of a non-parametric identification of the randomization-based standard error through a complete enumeration of all permutations ("permutation distribution") is limited for two main reasons:

1. While it is possible in principal to enumerate all possible permutations, their number becomes quickly prohibitively large. For example, when we randomly assign $n = 60$ subjects to two equal-sized groups of $k = 30$, we will obtain $\binom{60}{30} = 1.18 \cdot 10^{17}$ possible re-groupings. This is likely to exceed even the capacity of most modern computers (cf. Ramsey & Schafer, 2013: Chap. 1).

2. Determining the dispersion of the randomization distribution through the enumeration of all possible permutations requires the strict assumption that the treatment changes the outcome of *all* experimental subjects by an *identical* amount (of zero). Otherwise, we would be unable to determine individual counterfactuals for the unobserved treatment. Contrary to this unrealistic assumption, RCTs inherently allow for heterogeneous individual effects because the counterfactual is on the group level, and we can content ourselves with observing individual outcomes in only one of the two treatments. Therefore, we are interested in estimating an *average* treatment effect and its randomization-based standard error.

Usually, both problems can be solved by approximating the randomization-based standard error through the sampling-based standard error. The standard error of the difference between the mean of two randomized groups is quite similar to the standard error of the difference between the means of two groups in a random sample according to Eq. (3.4). To be more precise, Eq. (3.4) provides a convenient, while slightly upward-biased estimate of the standard error of the *average treatment effect* estimate obtained in a randomized design (Dunning, 2012: 186ff; Ramsey & Schafer, 2013: Chaps. 1 and 2; Athey & Imbens, 2017: 81).

If we accept this approximation of the randomization-based standard error, all other inferential statistics that researchers might possibly use, such as p-values or even statistical significance, can be simply based on the standard error of the sampling distribution. Of course, their drawbacks apply also in a randomization context: they represent an information downgrade compared to the two original pieces of information that we can derive from a random assignment, i.e., the causal effect estimate (signal) and the standard error (noise), which quantifies the uncertainty of the estimation caused by randomization error. Downgrading this information first into a quotient (such as a t- or z-ratio) and then into a p-value, and finally into a dichotomous significance declaration is, again, not by itself wrong. But it is rarely useful. It causes not only a substantial loss of information but virtually instigates overconfident causal conclusions, both when p is below and when it is above the "significance" threshold.[3]

7.2.3 Inference in Random Assignment of Randomly Recruited Subjects

Experiments based on randomization are primarily aimed at making inferences regarding average treatment effects in the given sample of experimental subjects (**sample average treatment effects**, SATE). However, generalizing causal inferences beyond the confines of a given sample toward a population (**population average treatment effects**, PATE) are often desired as well. Using statistical inference to help make such generalizations requires that, in addition to being randomly assigned to treatments, the experimental subjects are a random sample from a defined parent population. When we randomly recruit subjects from a population and then randomly assign them to different experimental treatments, the question arises of how to link randomization-based inference to sampling-based inference.

---

[3] Many experimental economists consider power analysis (cf. Sect. 6.3.2) as an important means to improve the quality of their research. But while the power concept originates from statistical decision theory according to Neyman and Pearson, experimental economists are usually not interested in deriving a behavioral rule under consideration of the error costs of two mutually exclusive decisions. Instead, they want to form reasonable beliefs regarding causal effects in the light of the estimation uncertainty caused by random error. Power indicates the repeatability of $p < 0.05$ under an ex ante specified alternative effect size hypothesis H_A. Therefore, criticisms of the dichotomy of NHST (cf. Sect. 6.2) apply to power analysis in equal measure. This holds unless its meaning is reduced to a rule of thumb that indicates how large experimental groups should be to produce, for a predefined alternative effect size and a presumed variance in the data, a "reasonably informative" signal-to-noise ratio (e.g., 1.645) in a certain share of statistical replications (e.g., 80%) defined by the power.

From a statistical point of view, the combination of both chance mechanisms represents an ideal case in which inferential statistical procedures can support causal inference regarding a treatment effect as well as generalizing inference toward the broader parent population. When there is both random sampling and randomization, the observed sample average treatment effect can be used as an estimate of the population average treatment effect. And we can use the sampling-based standard error of the mean difference according to Eq. (3.4) to assess the uncertainty of the estimation of the average treatment effect in the population.

While learning about causal effects in a parent population ("generalizing causal inference") is often desirable from a scientific point of view, many if not most experimental studies do not dispose of experimental subjects who are a random sample from a parent population. Therefore, it must not be forgotten that Eq. (3.4) is mostly used as a serviceable (upward-biased) approximation of the standard error that deals with uncertainty from randomization alone. When the experimental subjects are not a random sample, the interpretation of Eq. (3.4) must be strictly limited to causal inference *within* the *given* pool of experimental subjects. And researchers should explicitly communicate that all inferential statistics they report are solely concerned with the internal validity and uncertainty of the causal estimate (i.e., the sample average treatment effect) but no help for generalizing beyond the convenience pool of experimental subjects toward a broader population.

Confusion might arise because the mathematical formula for the standard error of a mean difference is identical in all three cases that facilitate statistical inference (cf. Table 7.1)—at least when we approximate the randomization-based standard error through the sampling-based standard error according to Eq. (3.4). Precisely because of this mathematical identity, researchers should always articulate the source(s) of uncertainty and stress that the meaning of the standard error depends on whether data were obtained through random sampling, randomization, or both. Only in the latter case, it can help make generalizing causal inferences by indicating the uncertainty of the estimation of a population average treatment effect.

7.3   Inferences Without Randomization or Random Sampling

Standard errors, p-values, and statistical significances are often displayed even when there is neither random sampling nor randomization. This is surprising because statistical inference is based on probability theory and therefore requires a probabilistic process of data generation. The problem seems to be that many researchers content themselves with routinely displaying p-values and making significance declarations without taking account of assumptions violations. Still worse, some researchers display p-values even without clarifying which kind of inference they aim at in the first place. Meaningful inferential reasoning requires being absolutely clear whether a statistical quantity such as a p-value, which is first of all a summary data statistic, is used as an auxiliary tool for drawing causal inferences, or for drawing generalizing inferences, or both. When support for causal inference is intended, the randomization procedure (or an imaginary equivalent) must be clearly described. And when support for generalizing inference is intended, the random sampling procedure (or an imaginary equivalent) as well as the parent population, from which the random sample was drawn and toward which inferences are to be made, must be clearly specified. Otherwise, displaying quantities such as standard errors, p-values, etc., is nonsensical, at best—and seriously misleading, at worst.

7.3.1   Fictitious Random Sampling

While often overlooked by those who routinely display p-values without a second thought, it is a logical necessity to *imagine* **a probabilistic process of data generation** ("chance mechanism") for inferential statistical procedures to make sense when there is neither random sampling nor randomization. While the usefulness of such an imagination is often questionable, the practice of more or less automatically displaying p-values whenever groups are compared is widespread. Therefore, we look into the question of how the imagination of a chance mechanism (fictitious random sampling) could possibly help make meaningful inferences. We illustrate the issue through two examples: a difference in mean outcomes between conveniently recruited male and female participants in a dictator game, and a difference in mean pay between a company's male and female employees who are considered to be the full population of interest. In both cases, we figure out the logical consequences that would result from the notion of fictitious random sampling.

Imagine the dictator game is carried out with a mixed-sex group of 100 haphazardly recruited participants. That is, we have a **convenience sample** but not a random sample of a broader population. Participants are asked to act as first players and decide which share of an initial endowment they want to transfer to an unknown second player. The amount they pass on is the only relevant outcome, and we are interested in the difference in mean outcomes between women and men. Due to engrained habits, researchers might be tempted to implement "statistical significance testing" routines in this dictator game even though there is no chance model upon which to base statistical inference. They would have to justify their routine by imagining **fictitious random sampling** from a **fictitious superpopulation**. This fictitious notion has already been critically discussed in Sect. 4.5. At this point, we therefore only summarize the essential arguments and relate them to the dictator game example:

1. No random sample was drawn from a numerically larger parent population. To justify inferential statistical procedures, one would therefore need to consider the convenience sample as the numerically finite, full population of interest.

2. To justify inferential statistical procedures, one would now need to imagine an infinite, unseen superpopulation, i.e., an underlying stochastic data generating process from which one happened to observe one random realization ("manifestation"). The convenience sample would thus become a random sample of the imagined superpopulation.
3. The difference in the mean transferred amount between the men and women in the convenience sample could now be interpreted as estimate of the unknown mean difference in the superpopulation. Accordingly, the sampling-based standard error for the mean difference according to Eq. (3.4) could be used as estimate of the standard deviation of a sampling distribution that would result from very frequent realizations of the underlying stochastic process.
4. The notion of a superpopulation cannot be used to support statistical inference in any conventional sense of the word. Instead of generalizing from a particular random sample toward a numerically larger parent population, "inferences" would be limited to the stochastic generating process that is presumably valid for *exactly* the 100 haphazardly recruited participants of the dictator game.

As already noted in Sect. 4.5, the notion of a superpopulation seems to be comprehensible to some and elusive to others. To make sense, one would have to equate the superpopulation, for example, with an underlying, quasi-stable real-world system whose operative regularities outlast the time interval over which the data were collected. In the dictator game example, this would imply that the unknown generating process not only produced the gender difference at the point in time when the experiment was carried out, but is at work for a prolonged period of time.

Even when we leave the question open of how much can be learned from the fiction of an imaginary superpopulation, one conclusion is unequivocal: *when* researchers resort to this fiction, they should clearly communicate *that* they use it. In this case, they must also transparently explain which kind of inferential reasoning the fiction-based standard error is meant to support. Without being transparent in this regard, confusion and misunderstandings are bound to arise. As already noted, this exigency is but a specification of the general desideratum that researchers clearly describe the sampling process and the population of interest from which the sample comes and to which they want to generalize.

Now, let us look at the announced observational study that investigates the difference in mean pay between male and female employees in a company. Imagine the company has a total of $n_f = 30$ female and $n_m = 30$ male employees. For all of them, payroll data are available. The notion of considering the available data as a random realization (**fictitious random sample**) of a stochastic generating process (**fictitious superpopulation**) would not differ from the one in the dictator game example. However, in the dictator game, we clearly had a convenience sample from a numerically larger population, but that sample had to be considered as the full population to justify statistical inference based on the notion of a superpopulation. Contrary to that, the company's employees are already the **full population** beyond which no inference toward a numerically larger population is intended.

The two examples illustrate that the comprehensibility and usefulness of imagining a superpopulation depends on the research context. Using the fiction makes generally more sense when we have data for a full population than when we only have a convenience sample of limited interest. This is because we may be genuinely interested in learning about a structural regularity (stochastic generating process, superpopulation) that underlies a particular data manifestation that we happened to observe in a population. In the company example, the superpopulation would have to be understood as the company's structural gender-related pay pattern—irrespective of its cause—that became manifest in the pattern of the studied payroll data. Consequently, the observed difference in mean pay would have to be understood as an unbiased point estimate (signal) of the company's structural gender pay gap. And the standard error according to Eq. (3.4) could be used to assess the uncertainty (noise) of this superpopulation estimate. Contrary to that, convenience samples are regularly not researchers' numerically final object of interest. This is why, in the dictator game example, the notion of a stochastic generating process that would be valid for exactly and only the 100 haphazardly recruited participants was so elusive and obviously of little interest.

The notion of a superpopulation in situations analogous to the gender pay gap example may be meaningful when the intended inference is clearly defined. But it cannot be emphasized enough that neither generalizing statistical inference toward a numerically larger population nor causal statistical inference regarding the effect of gender can be drawn. In an RCT, we would randomly assign subjects (with all their heterogeneous characteristics) to treatments, thus balancing confounders across groups and ensuring statistical independence of treatments. However, in the gender pay gap example, there was no random assignment—nor anything even slightly comparable. Therefore, the two gender groups are not statistically independent and inferential statistics cannot deal with the causal effect of gender on pay. A causal relationship may exist, but the difference could also be due to other reasons such as female employees working more part-time than male employees.

7.3.2   Fictitious Randomization

In real life, we can often observe subjects who were exposed to a treatment and others who were not. One of the many instances are government interventions that address specific social groups. A program granting investment subsidies to small-scale farmers with less land than a specified acreage would be an example. In policy impact evaluation, we are interested in establishing the causal effects of such programs. The problem is that we only have observational data. In other words, whether an individual subject belongs to the treatment group (i.e., receives the subsidy) or to the non-treatment group is *not* the result of random assignment. Instead, the (**non-random**) **assignment** is determined by other reasons, such as the objective of the intervention (here, improving the livelihood of small-scale farmers).

Random assignment would ensure that known *and* unknown confounders are balanced across both groups and, therefore, ex ante (by design) controlled for in the estimation of the average treatment effect. When we have only observations from non-randomly assigned groups, all attempts to ex post control for confounders through regression modeling are limited by the fact that confounding from *unobserved* variables cannot be controlled for in a model (Athey & Imbens, 2017). That is, the estimation of causal treatment effects based on observational data might be confounded by differences between the two groups other than the treatment.

Because of the relevance of establishing causal effects in contexts where treatments have not been assigned at random, a wide variety of "**quasi-experimental approaches**" (cf. Campbell & Stanley, 1963) have been suggested. These approaches, which are assumed to rehabilitate the probabilistic foundations for causal statistical inference, include matching, difference-in-differences, instrumental variables, and regression discontinuity designs (for an introduction, see Gertler et al., 2016). Recently, synthetic control methods (cf. Abadie & Gardeazabal, 2003; Abadie et al., 2010; Abadie, 2021) have also increasingly been used.

Quasi-experimental approaches are sophisticated attempts to move from correlation to causation by identifying a valid comparison group under consideration of the nature of the observational data. According to Shadish et al. (2002), three prerequisites are necessary for quasi-experimental approaches to allow a causal interpretation of relationships in observational data: (1) the cause precedes the effect, (2) the cause co-varies with the effect, and (3) alternative explanations for the causal relationship are implausible. If these conditions are met and if the move to causation is considered to be successful, we can use the standard error to assess the uncertainty of the estimation of a causal treatment effect. However, it must be observed that quasi-experimental approaches can only be used when the data allow us to use the notion of an imaginary equivalent of random assignment (**fictitious randomization**). To be more explicit, we must assume that, even though the treatment assignment was non-random, the average outcome of the non-treatment group represents an adequate (unbiased) estimate of the counterfactual, i.e., the average outcome that the treatment group would show without the treatment, and vice versa.

For illustration sake, let us look at **regression discontinuity designs** (RDD), which can be used when there is a continuous eligibility index and a clear eligibility threshold (cf. Gertler et al., 2016: 114). An early application by Thistlethwaite and Campbell (1960), which tackled the question of how merit awards influence academic outcomes, illustrates the basic principle. In the context under study, individuals win an award if their merit score exceeds a defined threshold. The idea of RDD is to use a "quasi-experimental sample" that consists only of the group of individuals *just* above the threshold (the winners of the award) and the group of individuals *just* below the threshold (the non-winners). Since there was no random assignment, using RDD requires arguing that the two groups are approximately equivalent (statistically independent), i.e., that potential outcomes and covariates are equally distributed in both groups. If we follow the argument that both groups are quasi identical except for the fact that the group just above the threshold received the award (treatment) and the other did not, the standard error can be used to estimate the uncertainty of the causal effect estimation. To be more precise, similar to an RCT, the standard error is presumed to be an adequate estimate of the standard deviation of the (fictitious) randomization distribution, i.e., the distribution of average treatment effects that would be observed over very frequently repeated *random* re-groupings of the subjects who are in the quasi-experimental sample under study.

To avoid inferential errors in research settings that seem to lend themselves to the estimation of causal effects via quasi-experimental approaches, researchers should keep in mind the following limitations:

- The trustworthiness of a **causal effect estimation based on observational data** depends on the ability of the quasi-experimental approach to identify a treatment and a non-treatment group that are statistically independent. The critical issue is whether the comparison group is able to produce an unbiased estimate of the counterfactual even though confounding from unobserved variables cannot be controlled for because the assignment was non-random. Only if the move to causation is trustworthy, the standard error can be interpreted as statistic that indicates the uncertainty of the causal effect estimation.
- Analogous to what holds for RCTs, causal statistical inference regarding an average treatment effect in a broader population (**generalizing causal statistical inference**) is only possible if the "quasi-experimental sample" was randomly recruited from that population. If this is not the case, the standard error is limited to dealing with the estimation uncertainty regarding the effect that the treatment causes in exactly and only the experimental sample under study.

7.4 Chapter Summary: Causal Inference Is Different from Generalization

With regard to the potential and limitations of causal statistical inference, several points should be kept in mind:

1. The informational content of the standard error is different in randomization-based inference as opposed to sampling-based inference. Randomization-based inference deals with internal validity and the uncertainty of causal inferences within the given group of experimental subjects. Sampling-based inference, in contrast, is concerned with external validity and the uncertainty of generalizing inferences from a sample to a population.
2. Random assignment—or an imaginary equivalent such as in quasi-experimental approaches—is necessary for a standard error to indicate the uncertainty of a causal effect estimation. If an experiment is not based on randomization or a plausible equivalent, causal inferences must be based on substantive arguments but cannot be supported by statistics. Examples are before-and-after-treatment comparisons where the validity of the comparison must be substantiated by research protocols that show what was done to hold everything but the treatment constant over time.

3. While being conceptually different, the sampling-based standard error can be used to approximate the uncertainty caused by randomization. But this approximation must not be confused with generalizing inference. When there was no random sampling from a population, generalizing inference requires substantive reasoning but cannot be supported by statistics. If an RCT is carried out with a convenience sample, statistical inference is limited to causal effect estimation in that sample (*sample* average treatment effect).

4. If one accepts the approximation and if the experimental subjects who participate in an RCT are randomly recruited from a population, the estimated average treatment effect can be generalized to the population; and the sampling-based standard error can be used to assess the uncertainty associated with the estimation of the *population* average treatment effect (generalizing causal inference).

5. In experiments, "control" means first of all avoiding that the comparison group differs from the treatment group in ways other than the treatment. Besides this internal validity issue, the term is also associated with external (ecological) validity. The expression "control over preferences" describes designs in which a valid measurement is achieved in that the behaviors of subjects in the experimental setting can be believed to reveal their true preferences. This is crucial for externally valid inferences, but it cannot be supported by inferential statistics.

7.5 Recommended Reading

Athey, S., & Imbens, G. W. (2017). The econometrics of randomized experiments. In A. V. Banerjee & E. Duflo (Eds.), *Handbook of field experiments* (Vol. 1, pp. 74–140). Elsevier.

Dunning, T. (2012). *Natural experiments in the social sciences: A design-based approach*. Cambridge University Press.

Gertler, P. J., Martinez, S., Premand, P., Rawlings, L. B., & Vermeersch, C. M. J. (2016). *Impact evaluation in practice* (2nd ed.). The World Bank.

Hirschauer, N., Grüner, S., Mußhoff, O., & Becker, C. (2020a). Inference in economic experiments. *Economics, The Open-Access, Open-Assessment E-Journal, 14*(2020-7), 1–14.

Imbens, G. W., & Rubin, D. B. (2015). *Causal inference for statistics, social, and biomedical sciences*. Cambridge University Press.

Ramsey, F. L., & Schafer, D. W. (2013). *The statistical sleuth: A course in the methods of data analysis*. Belmont.

Rubin, D. B. (2010): Causal inference. In P. Peterson, E. Baker, & B. McGaw (Eds.), International encyclopedia of education (3rd ed., pp. 66–71). https://www.sciencedirect.com/science/article/pii/B9780080448947013130

Shadish, W. R., Cook, T. D., & Campbell, D. T. (2002). *Experimental and quasi-experimental designs for generalized causal inference*. Houghton Mifflin.

Chapter 8
Better Inference in the 21st Century: A World Beyond $p < 0.05$

This book carved out the probabilistic assumptions and basic conception of statistical inference, in brief: random data generation, effect size estimation, and assessment of estimation uncertainty caused by random error. The book has been motivated by the fact that the concept of statistical inference is often reduced to so-called statistical significance testing, which is all-too-often ritualistically performed whenever there are quantitative data. Rather than helping make sound judgments regarding a study's knowledge contribution under consideration of random error, the ritual has caused considerable confusion and inferential errors.

The book provides researchers with a clear grasp of what statistical inference presupposes, what it can do, and what it cannot do. As regards content, what the book describes is not new. But being aimed at precluding inferential errors, it recalls the conceptual basics in a way that is, hopefully, so clear and impressive as to immunize researchers against the cognitive biases that are so intimately associated with dichotomous significance declarations. By focusing on the potential and the limits of statistical inference, the book is expected to make a contribution to answering the question of *what should be done to improve statistical inference in the 21st century* (cf. Wasserstein et al., 2019). The question of what should be done raises immediately another question: *who should bring about the changes which ensure that what should be done will be done*? A spontaneous but perhaps premature answer is that each researcher is individually responsible for an apt inferential reasoning of what one should reasonably believe regarding a real-world state of interest after seeing the results of a study.

It is certainly appropriate to hold the individual researcher responsible for following the rules of good scientific practice and avoiding inferential errors as best as possible: what should be done in the practice of research must be done by the individual researcher. However, an exclusive focus on the individual would miss the point if we did not also consider that research—and the use of statistics in research—is a complex social enterprise (Kasy, 2021). In this enterprise, the individual researcher, and especially a young researcher, is not the most potent initiator of reforms that are necessary to preclude damaging statistical reporting practices.

N. Hirschauer et al., *Fundamentals of Statistical Inference*, SpringerBriefs in Applied Statistics and Econometrics, https://doi.org/10.1007/978-3-030-99091-6_8

This is because those practices have been entrenched in the community for decades through inappropriate teaching, unwarranted reviewer requests, and even best-selling statistics textbooks. Instead, changes for the better depend to a large extent on institutions and their codes of conduct that govern the behavior of researchers. This includes, for example, statistical education, funding organizations, and professional associations. But above all, changes for the better depend on scientific journals with their methodological standards, guidelines, and review processes.

In the present system of research, peer-reviewed journals are—or should be—the gatekeepers of good scientific practice because they are key to what is published and thus publicly available as body of knowledge in any field of science. Therefore, the most urgent and decisive statistical reform is to revise journal guidelines and include explicit inferential reporting standards. Such standards should and could ensure that the "right inferential things" are not only done by researchers but also properly communicated and seen in published papers.

Seeing good inferential practices in publications is important because norms within scientific communities are established to a great deal through methodological mimicry. For example, when leading journals adopt new policies to improve the quality of published research, other journals are likely to follow suit and make it necessary for authors to comply with these new norms. The data-sharing policy adopted by the American Economic Association (AEA) in 2005, or its requirement to disclose conflicts of interest in 2012 are examples that brought about quick shifts to new norms (cf. Miguel, 2021). In contrast, inferential reporting policies by flagship journals were less successful in setting new standards. To combat misleading significance declarations, six of the most prestigious economics journals, *Econometrica*, the *American Economic Review,* and the four *American Economic Journals* have adopted guidelines that require authors to refrain from using asterisks or other symbols to denote statistical significance. Instead, they are asked to report the point estimates and standard errors. The reporting standard of the *American Economic Review* (AEA, 2021c), for example, reads as follows: "Do not use asterisks to denote significance of estimation results. Report the standard errors in parentheses." Unfortunately, this norm has not (yet) been widely adopted as best practice by other journals and national economic associations. On the contrary, many economists seem to be rather reluctant to abandon popular "asterisks-economics."

Given the model policy of those prestigious journals, other journals should follow suit and reform their guidelines based on the understanding that reporting point estimates and standard errors is a better summary of the evidence found in a dataset than statistical significance declarations. To improve inference, researchers must consider three essentials: first, statistical assumptions are empirical commitments, i.e., statistical inference presupposes a random process of data generation. Second, only in this case, it can help us infer something about a broader context based only on a limited dataset, such as *one* random sample drawn from a population, or *one* random assignment of experimental subjects to treatments. Third, even inferential tools that are more informative and less delusive than statistical significance are only auxiliary tools for inferential reasoning. Inferences are inherently uncertain and never flow from data-derived statistics automatically, whatever they may be.

The incorporation of inferential reporting standards into journal guidelines would bring about a consistent interplay of beneficial effects: first, it would effectively communicate necessary standards to researchers/authors. Second, it would help reviewers assess the credibility of inferential claims made in papers. Third, it would provide researchers/authors with an effective defense against unqualified reviewer demands. The protection from unwarranted reviewer request would arguably be even the most important benefit of including inferential reporting standards into journal guidelines. Such standards have the potential to considerably mitigate publication bias since many reviewers still seem to prefer statistically significant results and often pressure researchers to report p-values and "significant novel discoveries" without even taking account of whether data were randomly generated or not.

While the formulation and revision of journal guidelines is the responsibility of editors, including reporting standards in the spirit of the following suggestions is likely to make sense for many fields and journals. Dealing with causal statistical inference afterward, we, first of all, discuss suggestions that are concerned with generalizing statistical inference from samples to populations:

1. Take your data seriously and describe, first of all, the empirical evidence in terms of the **effect size** you found in your particular dataset (e.g., a sample mean, a sample mean difference, or a sample regression slope). Explicitly discuss whether this effect size is substantial and, for example, economically relevant.
2. Tackle inference in a separate, subsequent step and explicitly state whether you are concerned with **generalizing inference** or **causal inference**—or both.
3. When you are concerned with generalizing statistical inference, clearly define the **parent population** and, thus, the target population toward which you want to make inferences based on your sample.
4. Specify the **sampling design** (e.g., SRS, stratified sampling, cluster sampling) through which members of the population were selected into the sample. Remember: all statistics are derived from the data at hand. They only carry statistical inferential meaning when those data were randomly generated.
5. Clearly communicate that you want to use the effect size observed in your sample as a **point estimate** (**signal**) for the population effect size of interest. Estimate the **standard error** (**noise**) to assess the random-error-related uncertainty of that point estimation under explicit consideration of your sampling design (e.g., SRS, stratified sampling, cluster sampling).
6. Explicitly evaluate your study's contribution to the knowledge regarding your research question, under consideration of the point estimate (signal) and its uncertainty (noise) from random sampling error. In your **inferential reasoning**, consider that other types of error (e.g., measurement or specification error) may impair inferences much more than random sampling error.
7. *If* you should base your inferential reasoning on a signal-to-noise ratio (z- or t-score) or even a ***p*-value**, consider that downgrading the original signal and noise information first into a ratio and then into a p-value causes substantial losses of information. Therefore, a strong case must be made why a p-value (i.e.,

a graded measure of the compatibility of the data with the point null hypothesis) should be used as inferential indicator in your research setting.

8. Do not further downgrade the information and refrain from making dichotomous significance declarations based on p-value thresholds, i.e., **"retire statistical significance"** (cf. Amrhein et al., 2019; Wasserstein et al., 2019).

9. When you have **full population** data, do generally not display statistics such as standard errors or p-values because the uncertainty caused by random sampling error completely disappears when you have all data. *If* you nonetheless display inferential statistics, you face the logical necessity to base your reasoning on an imaginary **superpopulation**. In this case, explicitly communicate *that* you resort to this fiction and spell out *how* you define the superpopulation toward which you aim to generalize based on your full population data.

10. When you have a **convenience sample**, do not display inferential statistics such as standard errors or p-values, unless you have used a **"sample selection model"** that rehabilitates the probabilistic foundations for using inferential statistics. Remember: reporting standard errors or p-values in such a setting without correcting for selection bias would deceptively suggest that you can assess random sampling error even though there is no probabilistic model to do so.

For causal statistical inference in RCTs, corresponding reporting standards should be included into journal guidelines. Researchers/authors should also be asked to spell out that the standard error describes the uncertainty of the causal effect estimation caused by the vagaries of random assignment. In common settings where experimental subjects are not a random sample from a larger population, the standard error only quantifies the uncertainty of the causal effect estimation within the given sample of experimental subjects, i.e., the uncertainty associated with the estimation of the *sample* average treatment effect. However, it cannot be used for generalizations toward a population. This should be clearly articulated.

Using inferential statistics both for causal and generalizing conclusions ("generalizing causal inference") and assessing the uncertainty associated with the estimation of a *population* average treatment effect requires that experimental subjects are not only randomly assigned to treatments but randomly recruited from the parent population in the first place. In such a setting, researchers/authors should be asked to keep with the above reporting standards for generalizing inference. They should also be asked to articulate why and how their standard error estimates can be assumed to reflect the uncertainty associated with both causal and generalizing inference. Similar requirements could be made for quasi-experimental research aimed at estimating causal effects based on observational data.

Besides reporting standards in journal guidelines that are aimed at preventing inferential misinterpretations and miscommunications, the prevention of publication and reporting bias, including abuses such as p-hacking, is an additional field where reforms should be undertaken and continued. On the journal level, the most important step would be to *formally* inform reviewers that they should not discriminate against "non-significant" results or, to put it better, against studies with low signal-to-noise ratios. Such studies must equally be considered to obtain an appropriate picture of the population effect size of interest. Furthermore, preregistration and replication, which limit or, ideally, eliminate researchers' scope for cherry picking, should be promoted and established as a standard of good scientific practice whenever feasible. Reviewers' non-discrimination against "non-significant" results in conjunction with preregistration could considerably reduce inferential misapplications and publication bias, thus ensuring that the results of properly made studies see the light of the day irrespective of which results are found. The enterprise of science must be freed from systematic bias and made open-ended again.

References

Abadie, A. (2021). Using synthetic controls: Feasibility, data requirements, and methodological aspects. *Journal of Economic Literature, 59*(2), 391–425.

Abadie, A., Athey, S., Imbens, G. W., & Wooldridge, J. M. (2017). *When should you adjust standard errors for clustering?* NBER working paper 24003.

Abadie, A., Diamond, A., & Hainmueller, J. (2010). Synthetic control methods for comparative case studies: Estimating the effect of California's tobacco control program. *Journal of the American Statistical Association, 105*(490), 493–505.

Abadie, A., & Gardeazabal, J. (2003). The economic costs of conflict: A case study of the Basque country. *American Economic Review, 93*(1), 113–132.

AEA (2021a). *Data and code availability policy.* Accessed August 10, 2021, from https://www.aeaweb.org/journals/data/data-code-policy

AEA (2021b). *AEA RCT registry.* Accessed August 10, 2021, from https://www.socialscienceregistry.org/

AEA (2021c). *AER guidelines for accepted articles.* Accessed August 10, 2021, from https://www.aeaweb.org/journals/aer/styleguide#IV

Altman, N., & Krzywinski, M. (2017). Points of significance: P values and the search for significance. *Nature Methods, 14*(1), 3–4.

Amrhein, V., Greenland, S., & McShane, B. (2019). Retire statistical significance. *Nature, 567*, 305–307.

Andrews, I., & Kasy, M. (2019). Identification of and correction for publication bias. *American Economic Review, 109*(8), 2766–2794.

Athey, S., & Imbens, G. W. (2017). The econometrics of randomized experiments. In A. V. Banerjee & E. Duflo (Eds.), *Handbook of field experiments* (Vol. 1, pp. 74–140). Elsevier.

Auspurg, K., & Hinz, T. (2011). What fuels publication bias? Theoretical and empirical analyses of risk factors using the Caliper test. *Journal of Economics and Statistics, 231*(5–6), 636–660.

Austin, P. C. (2011). An introduction to propensity score methods for reducing the effects of confounding in observational studies. *Multivariate Behavioral Research, 46*, 399–424.

Becker, B. J., & Wu, M.-J. (2007). The synthesis of regression slopes in meta-analysis. *Statistical Science, 22*(3), 414–429.

Bedoya, G., Bittarello, L., Davis, J., & Mittag, N. (2017, July). *Distributional impact analysis. Toolkit and illustrations of impacts beyond the average treatment effect.* Policy research working paper WPS 8139. World Bank Group.

Berk, R. A. (1983). An introduction to sample selection bias in sociological data. *American Sociological Review, 48*(3), 386–398.

© The Author(s), under exclusive license to Springer Nature Switzerland AG 2022

N. Hirschauer et al., *Fundamentals of Statistical Inference*, SpringerBriefs in Applied Statistics and Econometrics, https://doi.org/10.1007/978-3-030-99091-6

Berk, R. A., & Freedman, D. A. (2003). Statistical assumptions as empirical commitments. In T. G. Blomberg & S. Cohen (Eds.), *Law, punishment, and social control: Essays in honor of sheldon messinger* (2nd ed., pp. 235–254). de Gruyter.

Berning, C., & Weiß, B. (2016). Publication bias in the German social sciences: An application of the Caliper test to three top-tier German social science journals. *Quality & Quantity, 50*(2), 901–917.

Berry, D. (2017). A p-value to die for. *Journal of the American Statistical Association, 112*(519), 895–897.

Borenstein, M., Hedges, L. V., Higgins, J. P. T., & Rothstein, H. R. (2009). *Introduction to meta-analysis*. Wiley.

Brick, J. M. (2013). Unit nonresponse and weighting adjustments: A critical review. *Journal of Official Statistics, 29*(3), 329–353.

Brodeur, A., Cook, N., & Heyes, A. (2020). Methods matter: P-hacking and publication bias in causal analysis in economics. *American Economic Review, 110*(11), 3634–3660.

Brodeur, A., Lé, M., Sangnier, M., & Zylberberg, Y. (2016). Star wars: The empirics strike back. *American Economic Journal: Applied Economics, 8*(1), 1–32.

Bruns, S. B. (2017). Meta-regression models and observational research. *Oxford Bulletin of Economics and Statistics, 79*(5), 637–653.

Camerer, C. F., et al. (2016). Evaluating replicability of laboratory experiments in economics. *Science, 351*, 1433–1436.

Cameron, A. C., & Miller, D. L. (2015). A practitioner's guide to cluster-robust inference. *Journal of Human Resources, 50*(2), 279–292.

Cameron, A. C., & Trivedi, P. K. (2005). *Microeconometrics: Methods and applications*. Cambridge University Press.

Cameron, A. C., & Trivedi, P. K. (2009). *Microeconometrics using Stata*. Stata Press.

Campbell, D. T., & Stanley, J. C. (1963). *Experimental and quasi-experimental designs for research*. Rand McNally.

Card, N. A. (2012). *Applied meta-analysis for social science research*. Guilford.

Center for Open Science. (2021). *Registered reports: Peer review before results are known to align scientific values and practices*. Accessed August 10, 2021, from https://www.cos.io/initiatives/registered-reports

Charness, G., Gneezy, U., & Kuhn, M. A. (2012). Experimental methods: Between-subject and within-subject design. *Journal of Economic Behavior & Organization, 81*(1), 1–8.

Cohen, J. (1994). The earth is round (p < 0.05). *American Psychologist, 49*(12), 997–1003.

Cooper, H., Hedges, L., & Valentine, J. (Eds.). (2009). *The handbook on research synthesis and meta-analysis* (2nd ed.). Russell Sage Foundation.

Copas, J. B., & Li, H. G. (1997). Inference for non-random samples. *Journal of the Royal Statistical Society Series B, 59*(1), 55–95.

Cuddeback, G., Wilson, E., Orm, J. G., & Combs-Orme, T. (2004). Detecting and statistically correcting sample selection bias. *Journal of Social Service Research, 30*(3), 19–33.

Czibor, E., Jimenez-Gomez, D., & List, J. A. (2019). *The dozen things experimental economists should do (more of)*. NBER working paper no. 25451.

Deeks, J. J., Higgins, J. P. T., & Altman, D. G. (2021). Chapter 10: Analysing data and undertaking meta-analyses. In J. P. T. Higgins, J. Thomas, J. Chandler, M. Cumpston, T. Li, M. J. Page, & V. A. Welch (Eds.), *Cochrane handbook for systematic reviews of interventions version 6.2* (updated February 2021). www.training.cochrane.org/handbook

De Leeuw, E. D., Hox, J. J., & Dillman, D. A. (2008). *International handbook of survey methodology*. Taylor & Francis.

Denton, F. T. (1988). The significance of significance: Rhetorical aspects of statistical hypothesis testing in economics. In A. Klamer, D. N. McCloskey, & R. M. Solow (Eds.), *The consequences of economic rhetoric* (pp. 163–193). Cambridge University Press.

Dreber, A., & Johannesson, M. (2019). Statistical significance and the replication crisis in the social sciences. In *Oxford research encyclopedia of economics and finance*. https://doi.org/10.1093/acrefore/9780190625979.013.461

Duflo, E., Glennerster, R., & Kremer, M. (2007). Using randomization in development economics research: A toolkit. In T. Schultz & J. Strauss (Eds.), *Handbook of development economics* (Vol. 4, pp. 3895–3962). Elsevier.

Dunning, T. (2012). *Natural experiments in the social sciences: A design-based approach*. Cambridge University Press.

Duvendack, M., Palmer-Jones, R., & Reed, W. R. (2015). Replications in economics: A progress report. *Econ Journal Watch, 12*(2), 164–191.

Duvendack, M., Palmer-Jones, R., & Reed, W. R. (2017). What is meant by "replication" and why does it encounter resistance in economics? *American Economic Review: Papers & Proceedings, 107*(5), 46–51.

Fisher, R. A. (1925). *Statistical methods for research workers*. Oliver and Boyd.

Fisher, R. A. (1935). *The design of experiments*. Oliver and Boyd.

Fisher, R. A. (1960). *The design of experiments* (7th ed.). Oliver and Boyd.

Foster, A., Karlan, D., Miguel, E., & Bogdanoski, A. (2019, July 15). *Pre-results review at the Journal of Development Economics: Lessons learned so far*. Development Impact. http://emiguel.econ.berkeley.edu/article/pre-results-review-at-the-journal-of-development-economics-lessons-learned-so-far/

Friedman, J. (2013, March 13). *Tools of the trade: When to use those sample weights worldbankblog*. https://blogs.worldbank.org/impactevaluations/tools-of-the-trade-when-to-use-those-sample-weights

Gelman, A. (2019). *Why do a within-person rather than a between-person experiment?* https://statmodeling.stat.columbia.edu/2019/11/16/why-do-a-within-person-rather-than-a-between-person-experiment/

Gerber, A. S., Green, D. P., & Kaplan, E. H. (2004). The illusion of learning from observational research. In I. Shapiro, R. Smith, & T. Massoud (Eds.), *Problems and methods in the study of politics* (pp. 251–273). Cambridge University Press.

Gerber, A. S., Malhotra, N., Dowling, C. M., & Doherty, D. (2010). Publication bias in two political behavior literatures. *American Politics Research, 38*(4), 591–613.

Gertler, P. J., Martinez, S., Premand, P., Rawlings, L. B., & Vermeersch, C. M. J. (2016). *Impact evaluation in practice* (2nd ed.). The World Bank.

Gigerenzer, G. (2004). Mindless statistics. *The Journal of Socio-Economics, 33*, 587–606.

Gigerenzer, G. (2018). Statistical rituals: The replication delusion and how we got there. *Advances in Methods and Practices in Psychological Science, 1*(2), 198–218.

Gigerenzer, G., & Marewski, J. N. (2015). Surrogate science: The idol of a universal method for statistical inference. *Journal of Management, 41*(2), 421–440.

Greenland, S., Senn, S. J., Rothman, K. J., Carlin, J. B., Poole, C., Goodman, S. N., & Altman, D. G. (2016). Statistical tests, P values, confidence intervals, and power: A guide to misinterpretations. *European Journal of Epidemiology, 31*(4), 337–350.

Guo, S., & Fraser, M. W. (2014). *Propensity score analysis. Statistical methods and applications* (2nd ed.). Sage.

Haller, H., & Krauss, S. (2002). Misinterpretations of significance: A problem students share with their teachers? *Methods of Psychological Research Online, 7*(1), 1–20.

Harrison, G. W., & List, J. A. (2004). Field experiments. *Journal of Economic Literature, 42*(4), 1009–1055.

Harvey, C. R. (2017). Presidential address: The scientific outlook in financial economics. *The Journal of Finance, LXXII*, 1399–1440.

Head, M. L., Holman, L., Lanfear, R., Kahn, A. T., & Jennions, M. D. (2015). The extent and consequences of P-hacking in science. *PLoS Biology, 13*(3), e1002106. https://doi.org/10.1371/journal.pbio.1002106

Heckman, J. J. (1979). Sample selection bias as a specification error. *Econometrica, 47*(1), 153–161.

Heckman, J. J., Ichimura, H., & Todd, P. E. (1997). Matching as an econometric evaluation estimator: Evidence from evaluating a job training programme. *Review of Economic Studies, 64*, 605–654.

Henrich, J., Heine, S. J., & Norenzayan, A. (2010). The weirdest people in the world? *Behavioral and Brain Sciences, 33*(2/3), 1–75.

Hirschauer, N. (2022). Unanswered questions in the *p*-value debate. *Significance, June*, 42–44.

Hirschauer, N., Mußhoff, O., Gruener, S., Frey, U., Theesfeld, I., & Wagner, P. (2016). Inferential misconceptions and replication crisis. *Journal of Epidemiology, Biostatistics, and Public Health, 13*(4), e12066-1–e12066-16.

Hirschauer, N., Grüner, S., Mußhoff, O., & Becker, C. (2018). Pitfalls of significance testing and *p*-value variability: An econometrics perspective. *Statistics Surveys, 12*, 136–172.

Hirschauer, N., Grüner, S., Mußhoff, O., & Becker, C. (2019). Twenty steps towards an adequate inferential interpretation of *p*-values in econometrics. *Journal of Economics and Statistics, 239*(4), 703–721.

Hirschauer, N., Grüner, S., Mußhoff, O., & Becker, C. (2020a). Inference in economic experiments. *Economics, The Open-Access, Open-Assessment E-Journal, 14*(2020-7), 1–14.

Hirschauer, N., Grüner, S., Mußhoff, O., Becker, C., & Jantsch, A. (2020b). Can *p*-values be meaningfully interpreted without random sampling? *Statistics Surveys, 14*, 71–91.

Hirschauer, N., Grüner, S., Mußhoff, O., & Becker, C. (2021a). A primer on p-value thresholds and α-levels – Two different kettles of fish. *German Journal of Agricultural Economics, 70*, 123–133.

Hirschauer, N., Grüner, S., Mußhoff, O., Becker, C., & Jantsch, A. (2021b). Inference using non-random samples? Stop right there! *Significance, October*, 20–24.

Hubbard, R., Haig, B. D., & Parsa, R. A. (2019). The limited role of formal statistical inference in scientific inference. *The American Statistician, 73*(Suppl 1), 91–98.

Hurlbert, S. H., & Lombardi, C. M. (2009). Final collapse of the Neyman-Pearson decision theoretic framework and rise of the neoFisherian. *Annales Zoologici Fennici, 46*, 311–349.

Imbens, G. W., & Rubin, D. B. (2015). *Causal inference for statistics, social, and biomedical sciences*. Cambridge University Press.

Imbens, G. W. (2021). Statistical Significance, *p*-values, and the reporting of uncertainty. *Journal of Economic Perspectives, 35*(3), 157–174.

Isaiah, A., & Kasy, M. (2019). Identification of and correction for publication bias. *American Economic Review, 109*(8), 2766–2794.

Jeffreys, H. (2011). *Scientific inference* (3rd ed.). Cambridge University Press.

Joober, R., Schmitz, N., Dipstat, L. A., & Boksa, P. (2012). Publication bias: What are the challenges and can they be overcome? *Journal of Psychiatry & Neuroscience, 37*(3), 149–152.

Kalaian, S. A., & Kasim, R. M. (2008). External validity. In P. J. Lavrakas (Ed.), *Encyclopedia of survey research methods* (pp. 254–257). Sage.

Kasy, M. (2021). Of forking paths and tied hands: Selective publication of findings, and what economists should do about it. *Journal of Economic Perspectives, 35*(3), 175–192.

Kennedy-Shaffer, L. (2019). Before p <0.05 to beyond p < 0.05: Using history to contextualize p-values and significance testing. *The American Statistician, 73*(Suppl 1), 82–90.

Kerr, N. L. (1998). HARKing: Hypothesizing after the results are known. *Personality and Social Psychology Review, 2*(3), 196–217.

Kish, L. (1965). *Survey sampling*. Wiley.

Kline, R. B. (2013). *Beyond significance testing: Statistics reform in the behavioral sciences*. American Psychological Association.

Knaub, J. (2008). Finite population correction (fcp) factor. In P. Lavrakas (Ed.), *Encyclopedia of survey research methods* (pp. 284–286). Sage.

Krämer, W. (2011). The cult of statistical significance – What economists should and should not do to make their data talk. *Schmollers Jahrbuch, 131*(3), 455–468.

Lavrakas, P. J. (2008). Internal validity. In P. J. Lavrakas (Ed.), *Encyclopedia of survey research methods* (pp. 345–351). Sage.

Leamer, E. E. (1978). *Specification searches: Ad hoc inference with nonexperimental data.* Wiley.

Lecoutre, B., & Poitevineau, J. (2014). *The significance test controversy revisited. The fiducial Bayesian alternative.* Springer.

Lehman, E. L. (1993). The Fisher, Neyman-Pearson theories of testing hypotheses: One theory or two? *Journal of the American Statistical Association, 88*, 1242–1249.

Levay, K. E., Freese, J., & Druckman, J. N. (2016). The demographic and political composition of mechanical turk samples. *SAGE Open, January–March*, 1–17. https://doi.org/10.1177/2158244016636433

Levy, P. S., & Lemeshow, S. (2008). *Sampling of populations. Methods and applications* (4th ed.). Wiley.

Little, R. J. A., & Rubin, D. B. (2020). *Statistical analysis with missing data* (3rd ed.). Wiley.

Lohr, S. L. (2019). *Sampling: Design and analysis* (2nd ed.). CRC Press.

Luellen, J. K., Shadish, W. R., & Clark, M. H. (2005). Propensity scores: An introduction and experimental test. *Evaluation Review, 29*, 530–558.

MacKinnon, J. G. (2019). *How cluster-robust inference is changing applied econometrics.* Queen's Economic Department, Working paper no. 1413.

McCloskey, D. N., & Ziliak, S. T. (1996). The standard error of regressions. *Journal of Economic Literature, 34*(1), 97–114.

McCulloch, C. E., Searle, S. R., & Neuhaus, J. M. (2008). *Generalized, linear, and mixed models* (2nd ed.). Wiley.

McGraw, K. O., & Wong, S. P. (1992). A common language effect size statistic. *Psychological Bulletin, 111*(2), 361–365.

Mercer, A. W., Kreuter, F., Keeter, S., & Stuart, E. (2017). Theory and practice in nonprobability surveys. Parallels between causal inference and survey inference. Special issue 2017: Survey research, today and tomorrow. *Public Opinion Quarterly, 81*(S1), 250–271.

Miguel, E. (2021). Evidence on research transparency in economics. *Journal of Economic Perspectives, 35*(3), 193–214.

Miller, P. V. (2017). Is there a future for surveys? Editorial to the special issue 2017: Survey research, today and tomorrow. *Public Opinion Quarterly, 81*(S1), 205–212.

Motulsky, J. J. (2014). Common misconceptions about data analysis and statistics. *The Journal of Pharmacology and Experimental Therapeutics, 351*(8), 200–205.

Munafò, M. R., Nosek, B. A., Bishop, D. V. M., Button, K. S., Chambers, C. D., du Sert, N. P., Simonsohn, U., Wagenmakers, E.-J., Ware, J. J., & Ioannidis, J. P. A. (2017). A manifesto for reproducible science. *Nature Human Behaviour, 1*(0021), 1–8.

National Academies of Sciences, Engineering, and Medicine. (2019). *Reproducibility and replicability in science. Consensus study report.* The National Academies Press. https://doi.org/10.17226/25303

Neyman, J., & Pearson, E. S. (1933a). On the problem of the most efficient tests of statistical hypotheses. *Philosophical Transactions of the Royal Society of London A, 231*, 289–337.

Neyman, J., & Pearson, E. S. (1933b). The testing of statistical hypotheses in relation to probabilities a priori. *Proceedings of the Cambridge Philosophical Society, 29*, 492–510.

Nickerson, R. S. (2000). Null hypothesis significance testing: A review of an old and continuing controversy. *Psychological Methods, 5*(2), 241–301.

Nosek, B. A., Ebersole, C. R., DeHaven, A. C., & Mellor, D. T. (2018). The preregistration revolution. *Proceedings of the National Academy of Sciences of the United States of America, 115*(11), 2600–2606.

Olken, B. A. (2015). Promises and perils of pre-analysis plans. *Journal of Economic Perspectives, 29*(3), 61–80.

Rabe-Hesketh, S., & Skrondal, A. (2021). *Multilevel and longitudinal modeling using Stata: Vols. I and II* (4th ed.). Stata Press.

Ramsey, F. L., & Schafer, D. W. (2013). *The statistical sleuth: A course in the methods of data analysis*. Belmont.

Roe, B. E., & Just, D. R. (2009). Internal and external validity in economics research: Tradeoffs between experiments, field experiments, natural experiments, and field data. *American Journal of Agricultural Economics, 91*(5), 1266–1271.

Rohatgi, V. K., & Saleh, A. K. M. E. (2015). *An introduction to probability and statistics* (3rd ed.). Wiley.

Rosenbaum, P. R. (2010). *Design of observational studies*. Springer.

Rosenthal, R. (1979). The file drawer problem and tolerance for null results. *Psychological Bulletin, 86*(3), 638–641.

Rosenthal, R., & Rosnow, R. L. (2009). *Artifacts in behavioral research*. Oxford University Press.

Rothstein, H., Sutton, A. J., & Borenstein, M. (2005). *Publication bias in meta-analysis. Prevention, assessment and adjustments*. Wiley.

Rubin, D. B. (2010): Causal inference. In P. Peterson, E. Baker, & B. McGaw (Eds.), International encyclopedia of education (3rd ed., pp. 66–71). https://www.sciencedirect.com/science/article/pii/B9780080448947013130

Schmidt, F. L., & Hunter, J. E. (2014). *Methods of meta-analysis: Correcting error and bias in research findings*. Sage.

Shadish, W. R., Cook, T. D., & Campbell, D. T. (2002). *Experimental and quasi-experimental designs for generalized causal inference*. Houghton Mifflin.

Simmons, J. P., Nelson, L. D., & Simonsohn, U. (2011). False-positive psychology: Undisclosed flexibility in data collection and analysis allows presenting anything as significant. *Psychological Science, 22*(11), 1359–1366.

Smith, M. L. (1980). Publication bias and meta-analysis. *Evaluation in Education, 4*, 22–24.

Smith, V. L. (1982). Microeconomic systems as an experimental science. *The American Economic Review, 72*(5), 923–955.

Sohn, D. (1998). Statistical significance and replicability: Why the former does not presage the latter. *Theory & Psychology, 8*, 291–311.

Song, F., Hooper, L., & Loke, Y. K. (2013). Publication bias: What is it? How do we measure it? How do we avoid it? *Open Access Journal of Clinical Trials, 5*, 71–81.

Stanley, T. D., & Jarrell, S. B. (1989). Meta-regression analysis: A quantitative method of literature surveys. *Journal of Economic Surveys, 3*(2), 161–170.

Sterling, T. D. (1959). Publication decisions and their possible effects on inferences drawn from tests of significance–Or vice versa. *Journal of the American Statistical Association, 54*(285), 30–34.

Thistlethwaite, D. L., & Campbell, D. T. (1960). Regression-discontinuity analysis: An alternative to the ex post facto experiment. *Journal of Educational Psychology, 51*(6), 309–317.

Trafimow, D. (2019). Five nonobvious changes in editorial practice for editors and reviewers to consider when evaluating submissions in a post $p < 0.05$ universe. *The American Statistician, 73*(Suppl 1), 340–345.

Trafimow, D., et al. (2018). Manipulating the alpha level cannot cure significance testing. *Frontiers in Psychology, 9*, 699. https://doi.org/10.3389/fpsyg.2018.00699

Valliant, R., Dever, J. A., & Kreuter, F. (2013). *Practical tools for designing and weighting survey samples*. Springer.

Vogt, W. P., Vogt, E. R., Gardner, D. C., & Haeffele, L. M. (2014). *Selecting the right analyses for your data: Quantitative, qualitative, and mixed methods*. The Guilford.

Wasserstein, R. L., & Lazar, N. A. (2016). The ASA's statement on *p*-values: Context, process, and purpose. *The American Statistician, 70*(2), 129–133.

Wasserstein, R. L., Schirm, A. L., & Lazar, N. A. (2019). Editorial: Moving to a world beyond "$p < 0.05$". *The American Statistician, 73*(Suppl 1), 1–19. [The full material of the special issue "Statistical Inference in the 21st Century: A World Beyond $p < 0.05$" can be accessed at: https://www.tandfonline.com/toc/utas20/73/sup1]

Weiß, B., & Wagner, M. (2011). The identification and prevention of publication bias in the social sciences and economics. *Journal of Economics and Statistics, 231*(5–6), 661–684.

Wooldridge, J. M. (2020). *Introductory econometrics: A modern approach* (7th ed.). Cengage.

Ziliak, S. T., & McCloskey, D. N. (2008). *The cult of statistical significance. How the standard error costs us jobs, justice, and lives.* The University of Michigan Press.

Index

© The Author(s), under exclusive license to Springer Nature Switzerland AG 2022
N. Hirschauer et al., *Fundamentals of Statistical Inference*, SpringerBriefs in Applied
Statistics and Econometrics, https://doi.org/10.1007/978-3-030-99091-6

Printed in the United States
by Baker & Taylor Publisher Services